SCIENCE AND TECHNOLOGY OF
PRINTING MATERIALS

SCIENCE AND TECHNOLOGY OF
PRINTING MATERIALS

Prakash Shetty

Additional Professor
Department of Chemistry
Manipal Institute of Technology
(A constituent Institute of Manipal University)
Manipal, Karnataka

MJP PUBLISHERS
Chennai 600 005

ISBN 10: 81-8094-044-6
ISBN 13: 978-81-8094-044-6

Cataloguing-in-Publication Data

Prakash Shetty (1963 –   ).
Science and Technology of Printing Materials / by
Prakash Shetty. – Chennai : MJP Publishers, 2008
xx, 386 p. ; 21 cm.
Includes Glossary, References and Index.
ISBN 81-8094-044-6 (pbk.)
1. Chemistry, other Organic Substances. I. Title.
547.8 PRA MJP 038

ISBN 81-8094-044-6   **MJP PUBLISHERS**
© Publishers, 2008   47, Nallathambi Street
All rights reserved   Triplicane
Printed and bound in India Chennai 600 005

Publisher : J.C. Pillai
Managing Editor : C. Sajeesh Kumar
Project Editor : P. Parvath Radha
Acquisitions Editors : M. Srinuvasan, C. Janarthanan
Assistant Editors : B. Ramalakshmi, S. Revathi
Composition : Lissy John, N. Yamuna Devi
Cover Designer : Lissy John
CIP Data : Prof. K. Hariharan

This book has been published in good faith that the work of the author is original. All efforts have been taken to make the material error-free. However, the author and publisher disclaim responsibility for any inadvertent errors.

To
My beloved Father

PREFACE

Materials Science has become a very important subject as an interdisciplinary course in almost all universities. Printing material science is one of the core subjects for Printing Engineering course both at the graduate and at the diploma level. The study of printing materials requires the basic knowledge of chemistry and its principles. It gives me great pleasure to put before the large community of Printing Engineering students, the first edition of my book *Science and Technology of Printing Materials*. This book particularly deals with the chemical materials used and chemical processes which are involved in graphic art operations. The chemical materials and processes play an important role in various areas of the graphic art industry. It is fundamental to the control of printing processes that the physical and chemical nature of materials should be understood. Simple and lucid explanations are adopted in the presentation of the topics. Most appropriate and illustrative diagrams have been included wherever necessary. This is an informational textbook prescribed for B.E. (Printing Engineering) course of various Universities and also for Diploma Course in Printing Engineering.

During the course of preparation of the manuscript I have gone through a number of reference books, journals and internet sites to make the book comprehensive and useful to the students. I take this opportunity to express my sincere thanks to the authors of these valuable works.

It is my proud privilege to thank Brig (Dr.) S. S. Pabla, Director, and Prof (Dr.) Girish Sharma, Head, Department of Print and Media Technology, Manipal Institute of Technology, Manipal, who have been the constant source of inspiration to my work. My special cordial thanks to Prof (Dr.) A. Nithyananda Shetty, Head, Department of Chemistry, NITK, Surathkal, for having taken pains to go through the manuscript and having made useful suggestions. I am indebted to Prof (Dr.) S. R. Girish, Head, Department of Chemistry and other colleagues who have directly or indirectly helped me in preparing this book.

I am grateful to MJP Publishers, Chennai, for their keen interest in the publication of this book with good production standards.

Any constructive criticism and suggestions from the valued readers for further improvement of the book will be gratefully accepted and acknowledged.

Prakash Shetty

CONTENTS

1 ACID-BASE CHEMISTRY

INTRODUCTION

The concepts of acids, bases, and salts are ancient ones that modern chemical science has adopted and refined. Our treatment of the subject at this stage will be mainly qualitative, emphasizing the definitions and the fundamental ideas associated with acids and bases.

ACIDS

The term "acid" was first used in the seventeenth century; it comes from the Latin root ac-, meaning "sharp" as in *acetum*, vinegar. Acids have long been recognized as a distinctive class of compounds whose aqueous solutions exhibit the following properties.

- characteristic sour taste
- ability to change the colour of litmus from blue to red
- react with certain metals to produce gaseous H_2
- react with bases to form a salt and water
- ability to conduct an electric current when dissolved in water
- increase the H^+ or H_3O^+ concentration in water

BASES

The term "base" has long been associated with a class of compounds whose aqueous solutions are characterized by the following properties.

- bitter taste
- a soapy feeling when applied to the skin
- ability to restore the original blue colour of litmus that has been turned red by acids
- ability to react with acids to form salts
- ability to conduct an electric current when dissolved in water
- decrease the H^+ or H_3O^+ concentration in water.

The word "alkali" is synonymous with base. It is of Arabic origin, but the root word comes from the same Latin *kalium* (potash) that is the origin of the symbol for potassium; wood ashes have been the traditional source of the strong base KOH since ancient times.

Neutralization Reaction

Acids and bases react with one another to yield two products: water and an ionic compound known as a salt. This kind of reaction is called neutralization reaction.

$$HCl + NaOH \longrightarrow NaCl + H_2O$$
$$H_2SO_4 + 2KOH \longrightarrow K_2SO_4 + 2H_2O$$

Both these reactions are exothermic. Although they involve different acids and bases, it has been determined experimentally that they all liberate the same amount of heat (57.7 kJ) per mole of H^+ neutralized. This implies that all neutralization reactions are really the one net reaction. This is given by the following reaction.

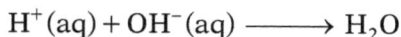

$$H^+(aq) + OH^-(aq) \longrightarrow H_2O$$

The "salt" that is produced in a neutralization reaction consists simply of the anions and cations that were already present. The salt can be recovered as a solid by evaporating the water.

ACID–BASE CONCEPTS

Different concepts have been put forth by various investigators to characterize acids and bases. Some of the important concepts of acids and bases are discussed below.

The Arrhenius Concept

According to the Arrhenius concept of acids and bases

- An acid is a substance that, when dissolved in water, increases the concentration of hydronium ions (H_3O^+).
- A base is a substance that, when dissolved in water, increases the concentration of hydroxide ions $OH^-(aq)$.

In the Arrhenius concept, a strong acid is a substance that ionizes completely in aqueous solution to give $H_3O^+(aq)$ and an anion. Chemists often use the notation $H^+(aq)$ for the H_3O^+ (aq) ion, and call it the hydrogen ion.

Examples   HCl, HBr, HI, $HClO_4$, HNO_3 and H_2SO_4.

Similarly, a strong base is a substance that ionizes completely in aqueous solution to give $OH^-(aq)$ and a cation.

Examples   NaOH LiOH, KOH, $Ca(OH)_2$, $Sr(OH)_2$ and $Ba(OH)_2$.

Most other acids and bases that we encounter are weak. They are not completely ionized and exist in reversible reaction with the corresponding ions. Some examples of weak acids are phosphoric acid (H_3PO_4), hydrofluoric acid (HF), acetic acid (CH_3COOH), oxalic acid ($H_2C_2O_4$), tartaric acid ($H_2C_4H_4O_6$), carbonic acid (H_2CO_3), etc.

A common example for a weak base is NH_4OH.

Bronsted–Lowry Concept

According to the Bronsted–Lowry concept of acids and bases

- An acid is the species that donates a proton in a proton-transfer reaction.
- A base is the species that accepts a proton in a proton-transfer reaction.

In any reversible acid–base reaction, both forward and reverse reactions involve proton transfer.

Consider the reaction of NH_3 and H_2O.

$$NH_3(aq) + H_2O \rightleftharpoons NH_4^+(aq) + OH^-(aq)$$

In the forward reaction, NH_3 accepts a proton from H_2O. Thus, NH_3 is a base and H_2O is an acid. In the reverse reaction, NH_4^+ donates a proton to OH^-. The NH_4^+ ion is the acid and OH^- is the base.

Water acts as an acid towards ammonia and as a base towards acetic acid. Such substances are said to be amphiprotic or amphoteric.

Conjugate acid–base pair   A conjugate acid–base pair consists of two species in an acid–base reaction, one acid and one base that differ by the loss or gain of a proton.

The species NH_4^+ and NH_3 are a conjugate acid–base pair.

$$NH_3(aq) + H_2O \rightleftharpoons NH_4^+(aq) + OH^-(aq)$$

From the above reaction, NH_4^+ is the conjugate acid of NH_3 and NH_3 is the conjugate base of NH_4^+. H_2O and OH^- is a conjugate acid–base pair.

The Lewis Concept

The Lewis definition of acids and bases states that a Lewis acid is an electron-pair acceptor, and a Lewis base is an electron-pair donor.

In the following Lewis acid–base reaction,

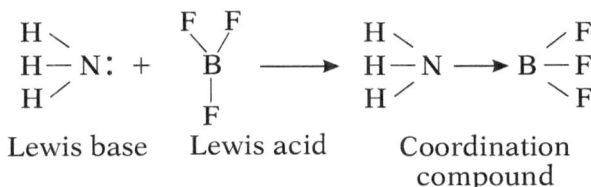

$$
\underset{\text{Lewis base}}{\begin{array}{c} H \diagdown \\ H-N: \\ H \diagup \end{array}} + \underset{\text{Lewis acid}}{\begin{array}{c} F \quad F \\ \diagdown / \\ B \\ | \\ F \end{array}} \longrightarrow \underset{\substack{\text{Coordination} \\ \text{compound}}}{\begin{array}{c} H \diagdown \qquad \diagup F \\ H-N \longrightarrow B-F \\ H \diagup \qquad \diagdown F \end{array}}
$$

Boron trifluoride (BF_3) molecule accepts a lone pair of electrons from ammonia (NH_3) molecule. Hence, BF_3 is a Lewis acid and NH_3 is a Lewis base. Thus a coordinate bond is formed between the Lewis acid and the Lewis base. (It is a special type of covalent bond in which the shared electrons are contributed only by one of the atoms involved in the bonding, and is usually represented by an arrow pointing away from the donor towards the acceptor.) The resulting combination is called a coordination compound.

Examples

Lewis acids: BF_3, SO_2, SO_3, CO_2, $AlCl_3$, BCl_3, Cu^{2+}, Zn^{2+}, etc.

Lewis bases: H_2O, NH_3, CO, NO, C_2H_4, C_2H_2, ROH, etc.

IONIC PRODUCT OF WATER

The ability of acids to react with bases depends on the tendency of hydrogen ions to combine with hydroxide ions to form water. This tendency is very great, so the reaction is practically complete. No reaction, however, is really 100 per cent complete. At equilibrium (when there is no further net change in the amount of substances), there will be at least a minute concentration of the reactants in the solution.

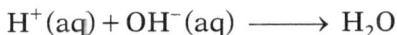

$$H^+(aq) + OH^-(aq) \longrightarrow H_2O$$

Another way of expressing this is to say that any reaction is at least slightly reversible.

This means that in pure water, the reaction,will proceed to a very slight extent.

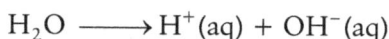

$$H_2O \longrightarrow H^+(aq) + OH^-(aq)$$

Experimental evidences confirm thus: the most highly purified water that chemists have been able to prepare will still conduct electricity very slightly. From this electrical conductivity, it can be calculated that the equilibrium concentration of both the H^+ and OH^- ions is almost exactly 1.0×10^{-7} mol/litre at 25°C. This amounts to one H_2O molecule in about 50 million being dissociated.

Thus, the product of the hydrogen ion and hydroxide ion concentrations in pure water or in any aqueous solution will always be 1.0×10^{-14} at 25°C.

In other words,

$$K_w = [H^+][OH^-] = 1.0 \times 10^{-14}$$

This expression is known as the ionic product of water, and it applies to all aqueous solutions, not just to the pure water. The consequences of this are far-reaching, because it implies that if the concentration of H^+ is large, that of OH^- will be small, and vice versa. This means that H^+ ions are present in all aqueous solutions, not just in acidic ones. This leads to the following important conclusions.

- In an acidic solution $[H^+] > [OH^-]$
- In an alkaline solution $[H^+] < [OH^-]$
- In a neutral solution $[H^+] = [OH^-]$ ($=1.0 \times 10^{-7}$ mol/litre at 25°C)

THE pH OF A SOLUTION

The possible values of $[H^+]$ and $[OH^-]$ in an aqueous solution can span many orders of magnitude, ranging from about 10^{-15} to 10 mol/litre. It is therefore convenient to represent them on a more compressed logarithmic scale. By convention, we use the pH scale. [This notation was devised by the Swedish Chemist Sorensen in 1909. The "p" as used in pH, pK, etc. stands for the German word Potenz which means "power" in the sense of an exponent to denote hydrogen ion concentrations.]

The pH of a solution is defined as the negative logarithm of the molar hydrogen-ion concentration.

$$pH = -\log_{10} [H^+]$$

Similarly, the pOH of a solution is defined as the negative logarithm of the molar hydroxide-ion concentration.

$$pOH = -\log_{10} [OH^-]$$

And for pure water or neutral solution at 25°C,

$$pH + pOH = 14$$

In a neutral solution at 25°C, the pH will be 7.0; a higher pH corresponds to an alkaline solution, a lower pH to an acidic solution. In a solution with $[H^+] = 0.0001$ M, the pH would be 4.0. Similarly, a 0.0001 M solution of NaOH would have a pOH of 4.0, and thus a pH of 10. It is very important to thoroughly understand the pH scale, and be able to convert between $[H^+]$ or $[OH^-]$ and pH in both directions.

- For neutral solutions (whose hydrogen-ion concentration is 1.0×10^{-7} mol/litre), the pH is equal to 7.00.
- For acidic solutions (whose hydrogen-ion concentration is greater than 1.0×10^{-7} mol/litre), the pH is less than 7.00.
- Similarly, for alkaline solutions (whose hydrogen-ion concentration is less than 1.0×10^{-7} mol/litre), the pH is greater than 7.00.

BUFFER SOLUTIONS

A buffer solution is the one which resists changes in the pH when small quantities of an acid or an alkali are added to it.

A buffer solution has to contain species which can remove any hydrogen ions or hydroxide ions that may be added to it. Otherwise, the pH will change. Acidic and alkaline buffer solutions achieve this in different ways.

Acidic Buffer Solutions

An acidic buffer solution is simply the one which has a pH less than 7. Acidic buffer solutions are commonly made from a weak acid and one of its salts (often a sodium salt).

A common example would be a mixture of acetic acid and sodium acetate in solution. In this case, if the solution contained equal molar concentrations of both the acid and the salt, it would have a pH of 4.76. It would not matter what the concentrations are as long as they are the same. The pH of the buffer solution can be changed by varying the ratio of acid to salt, or by choosing a different acid and one of its salts.

How do these buffer solutions work? Consider a mixture of acetic acid and sodium acetate as a typical example. Acetic acid is a weak acid and the position of this equilibrium will favour more to the left.

$$CH_3COOH \ (aq) \rightleftharpoons CH_3COO^-(aq) + H^+ \ (aq)$$

Adding sodium acetate to this, adds lots of extra acetate ions. According to Le Chatelier's principle, this addition will shift the position of the equilibrium even further to the left.

The effect of pressure, temperature and concentration on a system at equilibrium can be understood in a qualitative manner by the application of Le Chatelier's principle. It may be stated as follows, "if a change occurs in one of the factors influencing the equilibrium of a reversible reaction, the system will tend to adjust itself so as to nullify the effect of the change."

The solution will therefore contain these important things.

- plenty of non-ionized acetic acid
- plenty of acetate ions from the sodium acetate
- enough hydrogen ions to make the solution acidic

Adding an acid to this buffer solution The buffer solution must remove most of the new hydrogen ions otherwise the pH would drop markedly. Hydrogen ions combine with the acetate ions to form acetic acid. Although the reaction is reversible since the acetic acid is a weak acid, most of the new hydrogen ions are removed in this way.

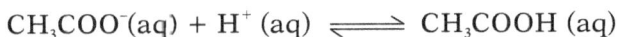

$$CH_3COO^-(aq) + H^+(aq) \rightleftharpoons CH_3COOH(aq)$$

Since most of the new hydrogen ions are removed, the pH does not change very much, but because of the equilibrium involved, it will fall a little bit.

Adding an alkali to this buffer solution Alkaline solutions contain hydroxide ions and the buffer solution must remove most of these. This time the situation is a bit more complicated because there are two processes which can remove the hydroxide ions.

Removal by reacting with acetic acid The most likely acidic substance with which a hydroxide ion is going to interact with is an acetic acid molecule. They will react to form acetate ions and water.

$$CH_3COOH(aq) + OH^-(aq) \rightleftharpoons CH_3COO^-(aq) + H_2O$$

As most of the new hydroxide ions are removed, the pH doesn't increase significantly.

Removal by reacting with hydrogen ions There are some hydrogen ions produced in the reaction given below.

$$CH_3COOH(aq) \rightleftharpoons CH_3COO^-(aq) + H^+(aq)$$

Hydroxide ions can combine with these to form water. As soon as this happens, the equilibrium shifts to replace them. This continues until most of the hydroxide ions are removed.

Again, as an equilibrium is involved, not all of the hydroxide ions are removed, just most of them. The water formed re-ionizes to a very small extent to give a few hydrogen ions and hydroxyl ions.

Alkaline Buffer Solutions

An alkaline buffer solution has a pH greater than 7. Alkaline buffer solutions are commonly made from a weak base and one of its salts.

A frequently used example is a mixture of ammonia solution and ammonium chloride solution. If these were mixed in equal molar proportions, the solution would have a pH of 9.25. Again, it does not matter what concentrations we choose as long as they are the same.

How do these buffer solutions work?   Consider a mixture of ammonia and ammonium chloride solutions as a typical example.

Ammonia is a weak base, and the position of this equilibrium will favour more to the left.

$$NH_3(aq) + H_2O \rightleftharpoons NH_4^+ (aq) + OH^-(aq)$$

Adding ammonium chloride to this, adds lots of extra ammonium ions. According to Le Chatelier's principle, that will shift the position of the equilibrium even further to the left.

The solution will therefore contain these important things.

● plenty of unreacted ammonia

● plenty of ammonium ions from the ammonium chloride

● enough hydroxide ions to make the solution alkaline

Adding an acid to this buffer solution   The two processes which can remove the hydrogen ions are as follows:

Removal by reacting with ammonia   The most likely basic substance with which a hydrogen ion is going to combine with is an ammonia molecule. They will react to form ammonium ions.

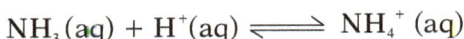

$$NH_3(aq) + H^+(aq) \rightleftharpoons NH_4^+(aq)$$

Most, but not all, of the hydrogen ions will be removed. The ammonium ion is weakly acidic, and so some of the hydrogen ions will be released again.

Removal by reacting with hydroxide ions   There are some hydroxide ions produced in the reaction between the ammonia and the water.

$$NH_3(aq) + H_2O \rightleftharpoons NH_4^+(aq) + OH^-(aq)$$

Hydrogen ions can combine with these hydroxide ions to make water. As soon as this happens, the equilibrium shifts to replace the hydroxide ions. This keeps on happening until most of the hydrogen ions are removed.

Again, because an equilibrium is involved, not all of the hydrogen ions are removed, just most of them.

Adding an alkali to this buffer solution   The hydroxide ions from the alkali are removed by a simple reaction with ammonium ions.

$$NH_4^+(aq) + OH^-(aq) \rightleftharpoons NH_3^+(aq) + H_2O$$

Because the ammonia formed is a weak base, it can react with water, and so the reaction is slightly reversible. Again, most (but not all) of the hydroxide ions are removed from the solution.

PREPARATION OF SOME IMPORTANT BUFFER SOLUTIONS

Buffer A   50 ml 0.2 M KCl + 0.2 M HCl (volume in ml indicated in Table).

Buffer B   100 ml 0.1 M potassium hydrogen phthalate + 0.1 M HCl (volume in ml indicated in Table).

Buffer C   100 ml 0.1 M potassium hydrogen phthalate + 0.1 M NaOH (volume in ml indicated in Table).

Buffer A		Buffer B		Buffer C	
pH	Volume added	pH	Volume added	pH	Volume added
1.00	134.0	2.20	99.0	4.10	2.6
1.10	105.6	2.30	91.6	4.20	6.0
1.20	85.0	2.40	84.4	4.30	9.4
1.30	67.2	2.50	77.6	4.40	13.2
1.40	53.2	2.60	70.8	4.50	17.4
1.50	41.4	2.70	64.2	4.60	22.2
1.60	32.4	2.80	57.8	4.70	27.2
1.70	26.0	2.90	51.4	4.80	33.0
1.80	20.4	3.00	44.6	4.90	38.8
1.90	16.2	3.10	37.6	5.00	45.2
2.00	13.0	3.20	31.4	5.10	51.0
2.10	10.2	3.30	25.8	5.20	57.6
2.20	7.8	3.40	20.8	5.30	63.2
		3.50	16.4	5.40	68.2
		3.60	12.6	5.50	73.2
		3.70	9.0	5.60	77.6
		3.80	5.8	5.70	81.2
		3.90	2.8	5.80	84.6
		4.00	0.2	5.90	87.4

Buffer D   100 ml 0.1 M KH_2PO_4 + 0.1 M NaOH (volume in ml indicated in Table).

Buffer E   100 ml 0.1 M tris (hydroxymethyl) aminomethane + 0.1 M HCl (volume in ml indicated in Table).

Buffer F   100 ml 0.025 M $Na_2B_4O_7 \cdot 10H_2O$ (borax) + 0.1 M HCl (volume in ml indicated in Table).

Buffer D		Buffer E		Buffer F	
pH	Volume added	pH	Volume added	pH	Volume added
5.80	7.2	7.00	93.2	8.00	41.0
5.90	9.2	7.10	91.4	8.10	39.4
6.00	11.2	7.20	89.4	8.20	37.6
6.10	13.6	7.30	86.8	8.30	35.4
6.20	16.2	7.40	84.0	8.40	33.2
6.30	19.4	7.50	80.6	8.50	30.4
6.40	23.2	7.60	77.0	8.60	27.0
6.50	27.8	7.70	73.2	8.70	23.2
6.60	32.8	7.80	69.0	8.80	19.2
6.70	38.6	7.90	64.0	8.90	14.2
6.80	44.8	8.00	58.4	9.00	9.2
6.90	51.8	8.10	52.4	9.10	4.0
7.00	58.2	8.20	45.8		
7.10	64.2	8.30	39.8		
7.20	69.4	8.40	34.4		
7.30	74.0	8.50	29.4		
7.40	78.2	8.60	24.4		
7.50	82.2	8.70	20.6		
7.60	85.6	8.80	17.0		
7.70	88.4	8.90	14.0		
7.80	90.6	9.00	11.4		
7.90	92.2				
8.00	93.4				

Buffer G 100 ml 0.025 M $Na_2B_4O_7 \cdot 10H_2O$ (borax) + 0.1 M NaOH (volume in ml indicated in Table).

Buffer H 100 ml 0.05 M $NaHCO_3$ + 0.1 M NaOH (volume in ml indicated in Table).

Buffer G		Buffer H	
pH	**Volume added**	**pH**	**Volume added**
9.20	1.8	9.60	10.0
9.30	7.2	9.70	12.4
9.40	12.4	9.80	15.2
9.50	17.6	9.90	18.2
9.60	22.2	10.00	21.4
9.70	26.2	10.10	24.4
9.80	30.0	10.20	27.6
9.90	33.4	10.30	30.4
10.00	36.6	10.40	33.0
10.10	39.0	10.50	35.6
10.20	41.0	10.60	38.2
10.30	42.6	10.70	40.4
10.40	44.2	10.80	42.4
10.50	45.4	10.90	44.0
10.60	46.6	11.00	45.4
10.70	47.6		
10.80	48.5		

Buffer I   100 ml 0.05 M Na_2HPO_4 + 0.1 M NaOH (volume in ml indicated in Table).

Buffer J   50 ml 0.2 M KCl + 0.2 M NaOH (volume in ml indicated in Table).

Buffer I		Buffer J	
pH	Volume added	pH	Volume added
10.90	6.6	12.00	12.0
11.00	8.2	12.10	16.0
11.10	10.2	12.20	20.4
11.20	12.6	12.30	25.6
11.30	15.2	12.40	32.4
11.40	18.2	12.50	40.8
11.50	22.2	12.60	51.2
11.60	27.0	12.70	64.4
11.70	32.4	12.80	82.4
11.80	38.8	12.90	106.0
11.90	46.0	13.00	132.0
12.00	53.8		

CALCULATION OF pH VALUES OF BUFFER SOLUTIONS

Acidic buffer solution   Consider a buffer solution containing a weak acid HA and its highly ionized salt NaA. The hydrogen ion concentration of the solution is given by the equation

$$[H^+] = K_a \frac{[Acid]}{[Salt]}$$

$$pH = pK_a + \log\frac{[Salt]}{[Acid]}$$

where,

pH $= -\log 10\ [H^+]$,

$pK_a = -\log_{10} K_a$ and

K_a is the dissociation constant of the acid.

This equation is known as Henderson's equation.

When the ratio [Salt] / [Acid] = 1, pH = pK_a. Since pK_a of the acid is constant at constant temperature, the pH of the buffer is also constant. Maximum buffering capacity is found when pH = pK_a, and buffer range is considered to be at a pH = $pK_a \pm 1$.

For example, pK_a for acetic acid is 4.75. Hence the buffer range for acetic acid–sodium acetate buffer is between pH 3.75 to 5.75.

Alkaline buffer solution   The pOH value of an alkaline buffer solution can also be calculated by Henderson's equation

$$pOH = pK_a + \log\frac{[Salt]}{[Base]}$$

where,

pOH $= -\log 10\ [OH^-]$,

$pK_b = -\log_{10} K_b$ and

K_b is the dissociation constant of the base.

The buffer action of an alkaline buffer solution is maximum when pH = $pK_w - pK_b$ (where K_w is the ionic product of water).

DETERMINATION OF pH VALUES OF SOLUTIONS

The common methods used to determine the pH of a solution are colorimetric method (using indicators) and electrometric method (using pH meter).

Colorimetric Method

Colorimetric method is based on the use of reagents which alter colour in accordance with the hydrogen ion concentration of solution. These reagents are weak organic acids or bases with undissociated molecules differing in colour from their ions, due to a difference in structure, and are commonly known as acid–base indicators. Some useful indicators for determining pH are complex organic compounds such as thymol blue, methyl orange, methyl red, phenol red, phenolphthalein, bromocresol green, bromophenol blue, etc. Each of these indicators changes colour over a range of about two pH units. Bromocresol green is yellow at a pH of 4.0 but changes through various shades of green until it is blue at a pH of 5.6. Such an indicator is useful only in this narrow range of pH. At any pH below 4.0, it remains yellow, and at any pH above 5.6 it remains blue. Measuring the pH of a solution outside of this range requires some other indicator. Thymol blue, which is sensitive to a pH range from 8.0 to 9.6, can be used in the pH range 8.0–9.6.

The test solution is taken in a test tube and then added a few drops of the selected indicator. The colour is compared with a series of control test tubes that contain the same indicator in solutions of different but known pH values. Such series of control tubes can be purchased for each indicator. The colour of the test solution are matched with that of one of the solutions in the series of test tubes, the pH is read as printed below the matching test tube.

Colorimetric determination of pH can be more conveniently carried out by using paper strips. These paper strips are impregnated with a solution of one of the indicators previously mentioned, and then dried. The paper strip is wetted with a drop of the test solution or immersed in the test solution. The colour developed in this paper strip is then compared with a pH colour chart that comes with the paper strips. The short-range pH papers of this kind are useful for an approximate determination of pH, such as the pH of a lithographic dampening solution.

Accuracy of determination can be improved by the use of so-called universal indicators, which are indicator mixtures changing colour over a wide range of pH.

Electrometric Method (Using pH Meter)

The pH of a given solution can be accurately measured using pH meter. The pH meter is associated with a cell for the measurement of pH. The cell consists of two electrodes. One is the glass electrode and the other is the saturated calomel electrode (a reference electrode). In practice, the glass electrode and the reference electrode are assembled into one unit that can be easily immersed into any solution whose pH is to be measured.

A pH meter will read pH very accurately if it is calibrated with a standard buffer solution that has a pH fairly close to the pH readings to be made. For example, if the pH readings are in the range of 3–5, one should use a standard buffer solution that has a pH of 4.0, or whatever is available that is close to this value. Assume that a standard buffer solution of pH 4.0 is selected. It is placed in a small beaker, and the combined electrode of pH meter is immersed in it. Using the calibration knob on the pH meter, the pH reading is adjusted until the pH meter reads exactly 4.0. The pH meter is now calibrated.

Once the pH meter is calibrated, determination of the pH of any solution is simple. The combined electrode of the pH meter is immersed in the test solution in a beaker. The meter shows the pH value of the test solution immediately. After each use, the electrode is washed with distilled water.

Technical Applications of pH Determination

In many industrial processes, a control and continuous recording of pH is most essential in order to carry out various operations in a satisfactory manner. Some applications of pH determination in industries are given below.

In paper industry   In the manufacture of paper, a proper control of pH is essential. In the paper industry, the sizing material (rosin soap) is precipitated by adding alum. Addition of small amount of alum can cause incomplete precipitation and an excess amount can lead to the corrosion of the equipment. In addition to this a low quality paper is obtained. The amount of alum added is checked by measuring the pH. For this the pH is controlled and maintained in between 4.0 to 6.0.

In printing industry   In the lithographic printing process, it is essential to control the pH of the fountain solution to obtain good plate performance and quality printing. The pH of acid fountain solution affects sensitivity, plate life, ink drying, etc. Ideal pH for most acid fountain solutions is in the range of 4.0 to 5.0. An accurate measurement of pH is possible using a pH meter and the pH is stabilized by adding buffering agent.

In electroplating and electrotyping   The pH of the electroplating bath must be properly controlled in the process of electroplating and electrotyping. This provides adherent, smooth and shining deposits. In nickel plating, the plating bath is maintained nearly at a pH of 5.5–5.8 by adding suitable buffers (boric acid). The deposits formed will be harder and brittle if the pH is less than 5.5. The deposit formed will be soft and dull if pH is greater than 5.8.

REVIEW QUESTIONS

1. Give the properties by which you can recognize a given solution as acid/base.

2. Discuss the Arrhenius concept of acids and bases.

3. Explain Bronsted–Lowry concept of acids and bases.

4. What is meant by a conjugated acid–base pair?

5. What are Lewis acids and bases?

6. Why is the reaction between an acid and a base called neutralization reaction?

7. Define pH of a solution.

8. What are buffer solutions? Explain the types of buffer solutions.

9. Explain the buffer action of an acidic and alkaline buffer solution.

10. Write the Henderson's equations used for calculating the pH value of an acidic and alkaline buffer.

11. Describe the colorimetric method for the determination of pH of a given solution.

12. Explain the method of determining the pH of a given solution using pH meter.

13. Discuss the technological applications of pH determination.

14. Explain the terms:
    i. Ionic product of water
    ii. Neutralization reaction

PROBLEMS

1. What is the pH of a buffer solution containing 0.2 mole of acetic acid and 0.2 mole of sodium acetate per litre. The dissociation constant of acetic acid is 1.8×10^{-5} at 25°C.

   (**Ans:** 3.74)

2. Calculate the pH of a solution with hydrogen ion concentration 5.2×10^{-4} mol/litre.

   (**Ans:** 3.28)

ADDITIONAL READING

Atkins, P. and Julio de Paula. (2002). *Atkins' Physical Chemistry*, 7th edn. Oxford University Press, New York.

Long, F.A. and Boyd, R.H. (1983). *Acids and Bases*: *McGraw-Hill Encyclopedia of Chemistry*. McGraw-Hill, New York.

Mahan, B.H. (1998). *University Chemistry*, 3rd edn. Narosa Publishing House, New Delhi.

Philip Mathews. (2003). *Advanced Chemistry (Physical and Industrial)*. Cambridge University Press.

2 WATER TECHNOLOGY

INTRODUCTION

Water is the commonest and the most important of all chemical compounds. It is essential for all forms of life. Moreover, water is used for all sorts of industrial processes, and as one of the best and cheapest solvents.

In the graphic art industry, water is widely used as solvent in the preparation of reagents for different processes. It is the main constituent of lithographic dampening solutions, photographic developers and fixing bath solutions. It is used for making light-sensitive coatings and developers for lithographic plates. Water is used as a solvent in flexographic and gravure inks for publications and package printing. The water being used in these applications should be reasonably pure. However, the natural waters from various sources usually contain varied amounts of different impurities. Hence, it is essential to check the purity of the incoming water in the printing and other related industries.

SOURCES OF WATER

The main sources of natural water are the following.

Rain water   It is the purest form of natural water since it is essentially distilled water. It contains traces of dissolved gases like oxygen, nitrogen, etc. and appreciable amounts of suspended particles.

River water   It contains salts in solution whose nature will vary with the ground over which the river flows. It also contains large amount of suspended impurities.

Well water   It is the water which has percolated large distances through the ground and become trapped between layers of non-porous rock. It usually contains noticeable quantities of dissolved salts and may be hard.

Sea water   This contains a high percentage (3.6%) of mineral salts.

IMPURITIES IN NATURAL WATER

Pure water is never found in nature. Natural water contains a variety of impurities in varying amounts. The following are the impurities commonly found in natural water.

Dissolved salts   Natural water contains soluble salts like chlorides, sulphates and nitrates in varying amounts. The presence of dissolved salts of calcium and magnesium causes hardness in water.

Dissolved gases   Water may contain many dissolved gases like oxygen, carbon dioxide, oxides of nitrogen and sulphur.

Suspended matter   These include fine particles of sand, clay and inorganic and organic suspended matter.

Microorganisms   Water containing organic impurities usually contains a large number of microorganisms.

HARDNESS OF WATER

Hard water is the water that does not easily give lather with soap. Soft water is the water that lathers easily with soap. The hard water fails to give lather because it contains dissolved salts of calcium and magnesium which react with the soap to give a curdy white precipitate. This is well described in the following reaction.

$$2RCOONa + Ca^{2+}(aq) \rightarrow (RCOO)_2Ca + 2Na^+$$

Sodium soap   Calcium soap
                                  (insoluble)

Types of Hardness

Temporary hardness (Carbonate hardness) It is caused by the bicarbonates of calcium and magnesium [$Ca(HCO_3)_2$ and $Mg(HCO_3)_2$] present in water. It is called "temporary" because it can be removed by boiling the water (when the bicarbonates decompose to give the precipitate of the respective carbonates).

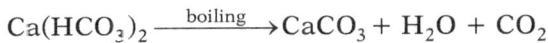

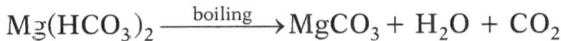

$$Ca(HCO_3)_2 \xrightarrow{\text{boiling}} CaCO_3 + H_2O + CO_2$$

$$Mg(HCO_3)_2 \xrightarrow{\text{boiling}} MgCO_3 + H_2O + CO_2$$

Permanent hardness (Non-carbonate hardness) It is usually caused by the presence of other soluble salts of calcium and magnesium in water. Unlike temporary hardness, permanent hardness cannot be removed by boiling. However, it can be removed by some special treatment methods.

The sum of temporary and permanent hardness is known as the total hardness of water.

Units of hardness Hardness of water is expressed in parts of calcium carbonate equivalents per million parts of water (ppm) or milligram of calcium carbonate equivalents per litre of water (mg/L).

SPECIFICATION OF WATER FOR PRINTING AND OTHER RELATED INDUSTRIES

The water that is used for various purposes in industries is commonly obtained from different sources like rivers, lakes and wells. It usually contains varying amounts of dissolved salts such as that of calcium, magnesium, iron and manganese. Hence the direct use of raw water in industries can cause many adverse effects. In the printing industry, water is extensively used in the preparation of dampening solutions, photographic developer and fixing bath solutions, lithoplate developers, deep-etch coating solutions, etc. In all these applications, it is essential to use reasonably pure water.

The water used for different industries has certain specifications. The specifications of water for printing and other related industries are given in Table 2.1.

Table 2.1   Specification of water for industries

Industries	Specification	Harmful effects
Printing	Reasonably soft	Increases the conductivity of dampening solutions and adversely affects the print quality.
	Free from salts of iron and manganese	Interferes with the normal working of various reagent solutions.
Paper	Absence of hardness and alkalinity	Calcium and magnesium salts increase the ash content of the paper. Consumption of alum by the alkaline water increases the cost of production.
	Free from salts of iron and manganese	Leads to discoloration of paper and affects the brightness of paper.
	Free from silica	Produces cracks in the paper.
Textile	Absence of hardness	Hard water precipitates basic dyes and decreases the solubility of dyes. Dyeing is not uniform.
	Free from salts of iron and manganese	Alters colours and shades of fabric.

PURIFICATION OF WATER

The impurities in water that are most important in the printing industry are discussed.

- Suspended matter, and
- Dissolved salts, particularly those causing hardness, i.e., salts of calcium and magnesium.

The suspended matter can be easily removed by filtration. Filtration is the process of removing suspended matter by passing water through a bed of fine sand and other proper-sized granular materials. It is carried out by using sand filters.

For some industrial uses, it is desirable to have nearly pure water, with most of the dissolved solids removed. The methods useful for this purpose are distillation, demineralization or deionization, and reverse osmosis. Deionization involves the removal of impurities from water, whereas distillation and reverse osmosis involve the removal of water from its impurities. The methods are futher discussed in detail.

Distillation

It is the simplest method to get pure or distilled water. The raw water on heating above the boiling temperature of water produces steam, which on cooling gives pure water. The condensate from industrial boilers also gives distilled water. The steam produced from the boiling water is almost free from all the mineral and organic matters that are present in raw water. The steam is then passed through condenser coils where it is cooled and converted back into liquid water called distilled water.

When water is boiling violently in a steam boiler, some tiny droplets of water containing dissolved salts are carried along with the steam. Hence, the condensate from industrial boilers may not give completely pure water. However, the degree of hardness will be reduced to a minimum level that is suitable for most applications. If still pure water is required, the distilled water can be distilled again and sometimes a third time. The energy costs are high for this method except when steam from an industrial boiler is available.

Deionization or Demineralization

The method of removal of all associated ions of water by the use of ion-exchange resins is known as deionization or demineralization of water. The synthetic ion-exchange resins are preferred to the naturally occurring complex aluminosilicate minerals, called zeolites.

Synthetic ion-exchange resins are usually the cross-linked organic polymers attached with ion-exchange groups. There are two types of ion-exchange resins—cation-exchange resins and anion-exchange resins. The cation-exchange resins exchange their H^+ ions with the other cations present in hard water. The anion-exchange resins exchange their OH^- ions with other anions present in water.

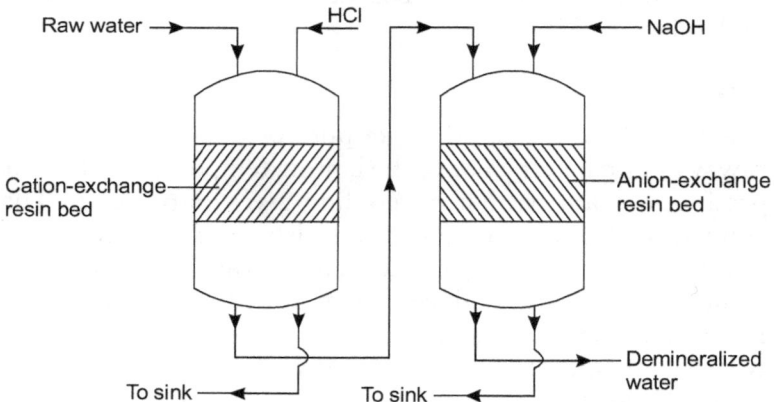

Figure 2.1   Demineralization of water

Process   The water is first passed through a cylindrical column loaded with cation-exchange resin bed (Figure 2.1). Here almost all of the cations in the water are exchanged with the H^+ ions of the resin. The exchange reaction may be represented as follows:

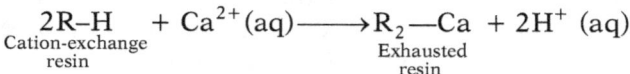

$$\underset{\substack{\text{Cation-exchange} \\ \text{resin}}}{2R–H} + Ca^{2+}(aq) \longrightarrow \underset{\substack{\text{Exhausted} \\ \text{resin}}}{R_2—Ca} + 2H^+ (aq)$$

The water coming out of the cation-exchange resin contains anions and OH^- ions. This water is now passed through a next column attached with an anion-exchange resin bed. Here almost all the anions in the water are exchanged with the OH^- ions of the resin. The exchange reaction may be represented as follows:

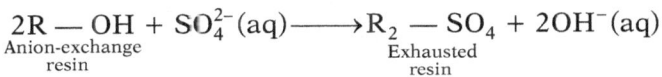

$$\underset{\substack{\text{Anion-exchange} \\ \text{resin}}}{2R - OH} + SO_4^{2-}(aq) \longrightarrow \underset{\substack{\text{Exhausted} \\ \text{resin}}}{R_2 - SO_4} + 2OH^-(aq)$$

The H^+ ions from cation resin combine with the OH^- ions from anion resin, to form water.

When the ion-exchange resins get exhausted, they can be regenerated by washing the cation-exchanger with a strong solution of hydrochloric acid or sulphuric acid and anion-exchanger with a strong solution of sodium hydroxide. The resin beds are then washed with deionized water and used again.

However, the un-ionized dissolved solids present in water cannot be removed by this method.

Reverse Osmosis

This is one of the methods used for the desalination of water. Desalination is the process of partial or complete removal of the dissolved salts from the seawater and brackish water. When two aqueous solutions of different concentrations are separated by a semi-permeable membrane, water passes through the membrane from the dilute solution to the concentrated solution. This process is known as osmosis and the flow of water is due to osmotic pressure. If a hydrostatic pressure in excess of osmotic pressure is applied on the concentrated solution, the flow of water is reversed. This process is known as reverse osmosis. The process of reverse osmosis can be used to separate pure water from the saline water/impure water. The saline water is separated from the pure water by a semi-permeable membrane which is permeable to water and not to the dissolved solids.

In the process, a pressure of about 15 to 40 kg cm^{-2} is applied on to the saline water. The pure water from saline water passes through the membrane, leaving behind the dissolved solids (both ionic as well as non-ionic). The membrane consists of very thin films of cellulose acetate, affixed to either side of a perforated tube. However, more recently superior membranes made of polymethacrylate and polyamide polymers have come into use.

A special type of reverse osmosis unit (Figure 2.2) consists of a series of porous tubes lined inside with thin film of cellulose acetate as semi-permeable membrane. These tubes are arranged in parallel in pure water. The saline water is pumped continuously at high pressure through these tubes. Pure water flows out from saline water through the membrane due to reverse osmosis. The rate of flow of purewater is proportional to the applied pressure which in turn depends on the nature of membrane used. Further, the greater the number of tubes, the larger the surface area and hence more production of pure water. The concentrated saline water and the pure water are continuously withdrawn through their respective outlets.

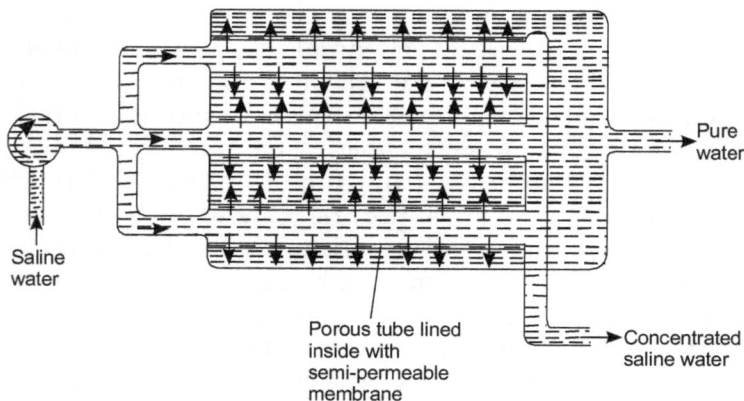

Figure 2.2   Reverse osmosis unit

Advantages

- It has a distinct advantage of removing ionic, non-ionic, colloidal and high-molecular-weight organic matters present in water.
- Requires low maintainence cost and hence it is more economical.
- The process needs extremely low energy.
- Unlike in demineralization, no regeneration is required.
- The lifetime of the membrane is quite high, about 2 years.

REVIEW QUESTIONS

1. What are the different types of impurities present in natural water?

2. Distinguish between hard water and soft water.

3. What are the salts in hard water interfering in the lathering of soap? Give the equations.

4. What do you mean by hardness of water? Explain the types of hardness.

5. Name the salts which are responsible for the temporary and permanent hardness of water.

6. How is the hardness of water expressed?

7. Explain the ion-exchange process for the removal of hardness from water.

8. Explain the disadvantages of using hard water in printing industry.

9. What is osmosis? How is reverse osmosis used for desalination of water?

10. What are ion-exchange resins? Discuss their applications in water-softening process. How are spent resins regenerated?

11. What is the principle of reverse osmosis? What is the main advantage of reverse osmosis over ion-exchange process?

12. Write short notes on:
    i. Reverse osmosis
    ii. Demineralization

13. What are the specifications for water to be used in the following industries?
    i. Paper industry
    ii. Printing industry
    iii. Textile industry

ADDITIONAL READING

Heitman, H.G. (1990). *Saline Water Processing.* VCH Publication, Weinheim, Germany.

Hoornaert, P. (1984). *Reverse Osmosis.* Pergamon Press, New York.

Khan, A.S. (1986). *Desalination Processes and Multi-stage Flash Distillation Practice.* Elsevier, Amsterdam, the Netherlands.

Spiegler, K.S. and Laird, A.D.K. (eds.). (1980). *Principles of Desalination,* 2nd edn. Academic Press, New York.

3

CHEMISTRY OF CARBON COMPOUNDS

INTRODUCTION

Most of the materials used in graphic arts are basically organic compounds. Thus, printing engineers or technicians, like the chemists, must have a fundamental knowledge of organic chemistry.

Today, organic chemistry is regarded as the chemistry of carbon compounds. In other words, organic chemistry deals with the compounds of carbon (except the inorganic substances such as carbon dioxide, carbon monoxide, carbonates, bicarbonates, and carbonic acid).

The essential components of all organic compounds are carbon and hydrogen. They may also contain oxygen, nitrogen, halogens, sulphur and phosphorus. The tetravalency of carbon and its unique property of forming long open- and closed chains give rise to a wide range of organic compounds.

Organic compounds, in general, differ greatly from inorganic compounds in their properties. Organic compounds have low melting point and boiling points, and are less soluble in water. The reactions of organic compounds are usually molecular rather than ionic. As a result, the reactions are often quite slow.

Organic compounds can be obtained from natural sources and by synthetic methods.

Natural sources are of three types.

Plant origin　Vegetable oils, cellulose, proteins, carbohydrates, gums, starch, sugar, etc. are obtained from plants.

Compounds like acetone, acetic acid, and methanol are obtained by destructive distillation of wood.

Animal origin   Proteins, animal oils and fats, vitamins, etc., are obtained from animals.

Mineral origin   Coal and petroleum are the two important examples. Coal-tar obtained from coal is the source of aromatic hydrocarbons, phenols, heterocyclic compounds, etc. Petroleum on fractionation gives a wide range of hydrocarbons, which serve as the starting material for the synthesis of many more organic compounds.

CLASSIFICATION OF ORGANIC COMPOUNDS

Organic compounds are broadly divided into two types: acyclic compounds and cyclic compounds.

Acyclic Compounds

Organic compounds having branched or unbranched open chain of carbon atoms are known as acyclic or aliphatic compounds. They are also known as open-chain compounds, for example, methane (CH_4), ethylene ($CH = CH$), acetone ($CH_3 — CO — CH_3$), etc.

Cyclic (or Ring) Compounds

Organic compounds having closed chain of carbon atoms are called cyclic or ring compounds. They may be monocyclic or polycyclic depending upon the number of rings associated with them. They are divided into two groups.

Homocyclic or carbocyclic compounds   They are the cyclic compounds having a ring of carbon atoms and are further divided into two groups.

Alicyclic compounds   Alicyclic compounds are the aliphatic cyclic compounds having a ring of carbon atoms. They resemble aliphatic hydrocarbons more than aromatic hydrocarbons.

Examples

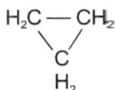

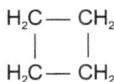

Cyclopropane   Cyclobutane   Cyclohexane

Aromatic hydrocarbons Cyclic compounds having a ring of six carbon atoms with alternate double bonds are called aromatic compounds. They may be monocyclic, bicyclic, or tricyclic, depending on whether they have one, two or three rings.

Examples

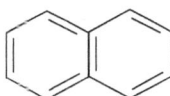

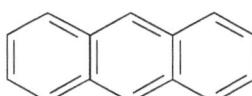

Benzene   Naphthalene   Anthracene

Heterocyclic compounds Cyclic compounds having a ring of carbon atoms with one or more heteroatoms like O, N, and S are called heterocyclic compounds.

Examples

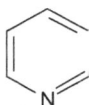

Pyridine   Furan   Thiophene

The above classification can be summarized as follows:

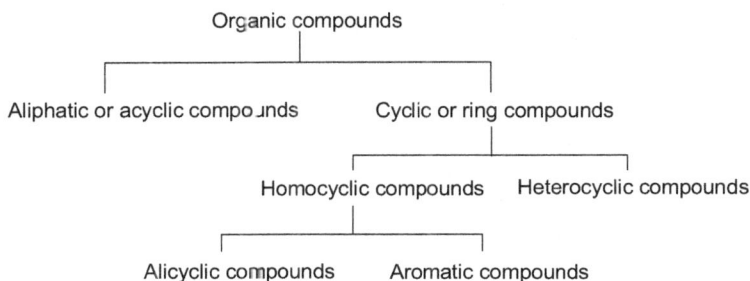

FUNCTIONAL GROUPS

Hydrocarbons are the backbones of organic compounds, as all other compounds are derived from them by the replacement of one or more hydrogen atoms with other atoms or groups. These groups are generally composed of heteroatoms like, O, N, S, halogens, etc. They are known as functional groups, since the properties of the compounds are largely governed by these groups.

A functional group is an atom or a group of atoms which defines the structure of a particular family of organic compounds and also determines their properties.

Some of the important functional groups are —Cl (chloro), —OH (hydroxyl), —SH (mercapto), —CHO (formyl), —CO— (carbonyl), —COOH (carboxyl), —COOR (ester), —NH$_2$ (amino), —N=N— (azo), —NO$_2$ (nitro), NO (nitroso), —SO$_3$H (sulphonic acid), —CN (cyano), —CONH$_2$ (amide) and —NCS (isothiocyanate).

HOMOLOGOUS SERIES

Organic compounds have been classified into a number of classes or homologous series. The characteristics of a homologous series are the following.

i. The members of a homologous series are represented by a general formula, for example, alkanes are represented by the general formula C_nH_{2n+2}.

ii. Every member of a homologous series has the same functional group and differs in molecular formula from the preceding and succeeding members by —CH$_2$ group. The individual member of a homologous series is called homologue.

iii. The members in any homologous series show different physical properties due to the difference in their molecular weight. Thus melting point and boiling point usually increase with an increase in molecular weight of the members in a homologous series.

iv. All members in any homologous series may be prepared by similar methods. They possess similar structures and similar chemical properties.

HYDROCARBONS

Organic compounds that contain only carbon and hydrogen are called hydrocarbons. The three main types of hydrocarbons are the aliphatic hydrocarbons, the aromatic hydrocarbons, and the alicyclic hydrocarbons.

Aliphatic (or acyclic) hydrocarbons They are the straight-chain or branched-chain hydrocarbons. Gasoline, kerosene, fuel oil and oils for heat-set inks consist largely of this type of hydrocarbons. They may be saturated or unsaturated hydrocarbons.

Examples Ethane, propane, isobutane, propene

Aromatic hydrocarbons They are the unsaturated hydrocarbons which contain at least one hexagonal six-carbon ring with three double bonds.

Examples Benzene, toluene, xylene

Alicyclic hydrocarbons They also have carbon atoms arranged in rings. They may be cycloparaffins (e.g. cyclobutane) or terpenes (e.g. turpentine).

Examples Cyclobutane, turpentine

SATURATED ALIPHATIC HYDROCARBONS (ALKANES)

Saturated aliphatic hydrocarbons are those hydrocarbons in which adjacent carbon atoms are joined by a single covalent bond and all other bonds are satisfied by hydrogen. They form a homologous series, called alkanes. They are quite inert towards most chemical reagents. For this reason they were termed "paraffins" by early chemists (Latin: *parum affinis* = little affinity). The members of the alkane series are represented by the general formula, C_nH_{2n+2} (where n is the total number of carbon atoms present).

The first three members of the alkane series are methane (CH_4), ethane ($H_3C—CH_3$) and propane ($H_3C—CH_2—CH_3$).

Compounds having same molecular formula but a different structural formula are known as isomers. Thus the two isomeric forms of butane are named as normal butane (n-butane) and isobutane.

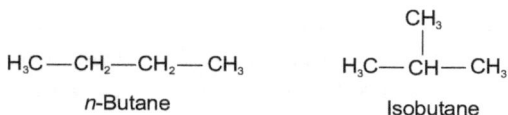

$$H_3C—CH_2—CH_2—CH_3 \qquad\qquad H_3C—\overset{\overset{\displaystyle CH_3}{|}}{C}H—CH_3$$

n-Butane   Isobutane

Similarly, the three isomers of pentane are n-pentane, isopentane and neopentane.

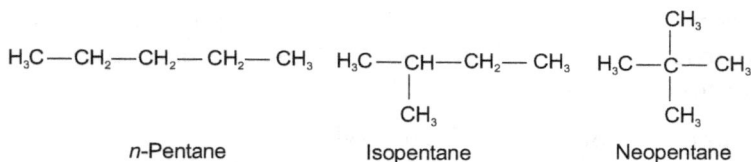

$$H_3C—CH_2—CH_2—CH_2—CH_3 \quad H_3C—\underset{\underset{\displaystyle CH_3}{|}}{C}H—CH_2—CH_3 \quad H_3C—\overset{\overset{\displaystyle CH_3}{|}}{\underset{\underset{\displaystyle CH_3}{|}}{C}}—CH_3$$

n-Pentane   Isopentane   Neopentane

As the number of carbon atoms increases in the molecule, the number of possible isomers increases accordingly. Thus, there are five possible isomers of hexane (C_6H_{14}) and 75 possible isomers of decane ($C_{10}H_{22}$).

The saturated aliphatic hydrocarbons are colourless, practically odourless, and quite insoluble in water, particularly those with five or more carbon atoms. They dissolve readily in many organic solvents. At room temperature all members through C_5 are gases, those from C_6 to C_{17} are liquids and those above C_{17} are solids.

The alkane series is characterized by names ending in "ane". The first four members of this series are known as methane (CH_4), ethane (C_2H_6), propane (C_3H_8) and butane (C_4H_{10}). Rest of the members are named by prefixing Greek numeral, indicating the number of carbon atoms in the molecule, to the suffix "ane". Thus pentane, hexane, heptane, etc. represent the members with molecular formulae, C_5H_{12}, C_6H_{14}, C_7H_{16}, etc.

The saturated hydrocarbons have four different types of carbon atoms in their molecules.

    i. *Primary carbon atom* The carbon atom which is attached either to no other carbon atoms or to only one other carbon atom.

    ii. *Secondary carbon atom* The carbon atom which is attached to two other carbon atoms.

    iii. *Tertiary carbon atom* The carbon atom which is attached to three other carbon atoms.

    iv. *Quaternary carbon atom* The carbon atom which is attached to four other carbon atoms.

Straight-chain alkanes contain only primary and secondary carbon atoms and are termed as normal (abbreviated as 'n') hydrocarbons (e.g. *n*-pentane). The branched-chain alkane having a secondary carbon atom is called isoalkane (e.g. isopentane). Similarly, the one with a quaternary carbon atom is called neoalkane (e.g. neopentane).

The IUPAC (International Union of Pure and Applied Chemistry) name of saturated hydrocarbon is alkane. In this system, the common names of alkanes have been retained for the normal-chain hydrocarbons. However, the branched-chain hydrocarbons are named by selecting the longest continuous chain of carbon atoms as the parent chain. The carbon atoms in the chain are then numbered from one end to the other by Arabic numerals, in a manner such that the carbon atoms having substituents get the lowest number. The side chains are indicated by the number of carbon atoms to which these are attached and prefixed to the name of parent alkane.

Examples

$$CH_3—CH—CH_2—CH_3$$
$$|$$
$$CH_3$$

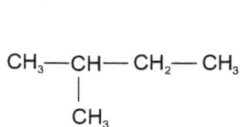

2-methyl butane

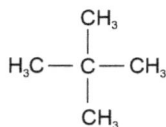

$$CH_3$$
$$|$$
$$H_3C—C—CH_3$$
$$|$$
$$CH_3$$

2,2'-dimethyl propane

The alkanes are the main constituents of petroleum and natural gas. Petroleum is a mixture of many of the long-chain saturated hydrocarbons. Natural gas from petroleum wells contains methane, ethane, propane and butane. Methane is produced during the natural decay of animals and plants.

UNSATURATED ALIPHATIC HYDROCARBONS

Aliphatic hydrocarbons which contain a double or triple bond between adjacent carbon atoms in the chain are known as unsaturated aliphatic hydrocarbons. Unsaturated hydrocarbons with a carbon-to-carbon double bond are known as alkenes or olefins. Those with carbon-to-carbon triple bonds are known as alkynes or acetylenes.

Alkenes

The members of these homologous series are represented by the general formula "C_nH_{2n}". They have two hydrogen atoms less than the corresponding alkanes. They are also known as olefins because lower gaseous members of the series react with chlorine to form oily products. The simplest olefin has the formula C_2H_4 and is commonly known as ethene or ethylene.

The common names of the alkenes are derived from the corresponding alkanes by replacement of "ane" of the common name of the alkane by "ene" (or ylene). In naming specific alkenes, the IUPAC system must be employed on all compounds with over three carbon atoms. The parent alkene is considered to be the one corresponding to the longest chain of carbon atoms containing the double bond. The carbon chain is then numbered from the end nearer to the double bond and its position is indicated by the number of the carbon atoms preceding the double bond. The nomenclature becomes quite complicated with branched-chain isomers.

Examples

$$CH_2 = CH_2 \qquad CH_3 - CH = CH_2 \qquad CH_2 = CH - CH_2 - CH_3$$

Ethene   Propene   1-Butene

$$CH_3 - CH = CH - CH_3 \qquad CH_3 - CH = CH_2 \qquad CH_3 - CH_2 - CH_2 = CH_2$$
$$\qquad\qquad\qquad\qquad\qquad\qquad\quad | \qquad\qquad\qquad\qquad\qquad |$$
$$\qquad\qquad\qquad\qquad\qquad\qquad CH_3 \qquad\qquad\qquad\qquad\qquad CH_3$$

2-Butene   2-Methyl-1-propene   2-Methyl-1-butene

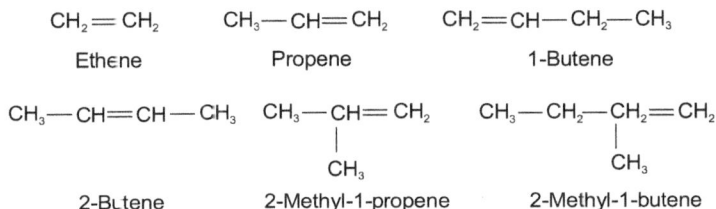

Lower alkenes occur in coal gas but in very small amounts. Ethene is present in natural gas. They are also produced as products of destructive distillation of organic materials. First three members of alkene series (ethene, propene and butene) are gases at room temperature, while alkenes having five to fifteen carbon atoms are liquids and the higher members are solids. They are insoluble in water but are soluble in certain organic solvents like alcohol. They have characteristic smell and burn with luminous flame. They are less volatile than alkanes. Their boiling point and melting point, and specific gravity are higher than corresponding alkanes.

Alkenes are more reactive than alkanes. Under proper conditions, ethylene will react with hydrogen, bromine, hydrochloric acid, etc. Ethylene also undergoes long-chain reactions called polymerization and this results in the plastic called polyethylene. A lot of printing, particularly by flexography, is done on film of polyethylene.

Alkynes (or Acetylenes)

The members of this homologous series are represented by the general formula, C_nH_{2n-2}. They are characterized by a carbon-to-carbon triple bond in their chain. The first member of the alkyne series has the formula, C_2H_2, and is known as acetylene. Therefore, alkynes are also known by the general name acetylenes.

In the IUPAC system of nomenclature of alkynes the ending "ane" of the corresponding alkane is replaced by "yne". The parent carbon chain is numbered in such a way so as to assign minimum number to the carbon atom preceding the triple bond.

Examples

$$CH\equiv CH \qquad CH_3-C\equiv CH \qquad CH\equiv C-CH_2-CH_3$$

Ethyne   Propyne   1-Butyne

$$CH_3-C\equiv C-CH_3 \qquad\qquad CH_3-CH-C\equiv CH$$
$$\underset{\displaystyle CH_3}{|}$$

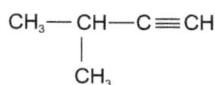

2-Butyne   3-Methyl-1-butyne

Alkynes do not occur in the nature in the free-state because of their higher reactivity. However, they are produced by combustion of coal and decomposition of organic matter. Large amounts of alkynes, particularly acetylene, are obtained by cracking of various fractions of petroleum.

The lower alkynes up to four carbon atoms are gases, those containing five to thirteen carbon atoms are liquids and higher alkynes are solids. They are colourless and odourless compounds (except acetylene which has garlic odour). They are sparingly soluble in water but are readily soluble in organic solvents such as ethanol, acetone, and benzene. Lower alkynes produce general anaesthesia, when inhaled in large amounts.

ALKYL GROUPS

The removal of one hydrogen atom in any alkane results in a group, known generally as alkyl group or organic group. Their names, which usually end in "yl", are derived from the names of the corresponding alkanes. These alkyl groups are generally represented by the symbol "R" and have the general formula, C_nH_{2n+1}.

Alkane	Alkyl group
CH_4 (methane)	CH_3— (methyl)
C_2H_6 (ethane)	C_2H_5— (ethyl)
C_3H_8 (propane)	C_3H_7— (propyl)
C_4H_{10} (butane)	C_4H_9— (butyl)

The group like this cannot exist alone, but they do occur in many organic compounds in combination with some other group, called functional group. The names of the alkyl groups are often used in the naming of an organic compound.

Alkyl Halides

Organic compounds derived from alkanes by the replacement of one or more hydrogen atoms by the halogen atoms (fluorine, chlorine, bromine or iodine) are known as alkyl halides (or haloalkanes). The general notation commonly used for representing alkyl halides is R–X, where R stands for an alkyl group and X for any halogen.

The simplest alkane, methane (CH_4), forms four alkyl halides by the replacement of one, two, three or all the four hydrogen atoms of methane. Thus chlorination of methane gives: CH_3Cl (methyl chloride or chloromethane), CH_2Cl_2 (methylene chloride or dichloromethane), $CHCl_3$ (chloroform or trichloromethane) and CCl_4 (carbon tetrachloride or tetrachloromethane).

Based on the number of halogen atoms present in an alkyl halide, they are called as mono-, di-, tri-, tetra-, and polyhalogen derivatives.

The common names of the alkyl halides are derived from those of the corresponding hydrocarbons and they are named as if they are the alkyl derivatives of hydrogen halides. Thus $CHCl_3$ may be considered as chloride of the methyl group and is named methyl chloride. Usually the name of the alkyl group precedes the name of halogen, e.g. CH_3Br (methyl bromide), C_2H_5Cl (ethyl chloride), etc.

Alkyl halides are subdivided into primary, secondary and tertiary depending upon the nature of carbon atom to which halogen is attached.

Primary alkyl halides CH_3Cl (methyl chloride), C_2H_5Cl (ethyl chloride).

Secondary alkyl halides CH_3—CH Br—CH_3 (isopropyl bromide), CH_3—CH_2—CH Br—CH_3 (sec-butyl bromide).

Tertiary alkyl halide $(CH_3)_3$—C.Br (tert-butyl bromide).

In IUPAC system alkyl halides are named as halogen derivatives of alkanes. The name of halogen is prefixed to the name of alkane and the name of alkyl halide is written in one word, for example: CH_3I (iodomethane), CH_3—CH Br—CH_3 (2-bromopropane), $(CH_3)_3$—CI (2-iodo-2-methyl propane).

The lower members of alkyl halides (like methyl chloride, methyl bromide and ethyl chloride) are colourless gases while some of the higher members are colourless sweet-smelling liquids but still higher homologues are odourless solids. They are generally insoluble in water and are soluble in organic solvents (like alcohol, ether, etc). Alkyl halides are in general toxic compounds and cause general anaesthesia when inhaled in large quantities.

Alcohols

Alcohols may be considered as hydroxyl derivatives of hydrocarbons obtained by replacing one or more hydrogen atoms of a hydrocarbon by hydroxyl (—OH) groups. Alcohols containing only one hydroxyl group are known as monohydric alcohols. They may be saturated or unsaturated depending on the hydrocarbon group present. The saturated monohydric alcohols form a homologous series having the general formula, $C_nH_{2n+1}OH$. They are further classified into three groups as primary, secondary and tertiary alcohols.

If hydroxyl group is attached to a primary carbon atom, i.e., a terminal carbon atom, the alcohol is termed as primary alcohol, e.g. CH_3CH_2OH (ethyl alcohol), $CH_3CH_2CH_2OH$ (propyl alcohol).

If the hydroxyl group is attached to a secondary carbon atom, i.e., a carbon atom that is attached to two other carbon atoms, the alcohol is known as secondary alcohol, e.g. $CH_3CH(OH)CH_3$ (isopropyl alcohol), $CH_3CH_2CH(OH)CH_3$ (sec-butyl alcohol).

If the hydroxyl group is attached to a carbon atom that is attached to three other carbon atoms (i.e., tertiary carbon atom), then it is known as tertiary alcohol, e.g. $(CH_3)_3.C.OH$ (tert-butyl alcohol).

The simpler alcohols are named after the alkyl group to which the hydroxyl group is attached, e.g. CH_3OH is known as methyl alcohol. The IUPAC system must be employed, however, to differentiate among isomers and to name the higher members. In IUPAC system alcohols are named as alkanols. Thus, C_2H_5OH is known as ethanol. In this system, the longest carbon chain containing the hydroxyl group determines the name. The location of alkyl groups and the hydroxyl group are described by number.

Structural formula	Common name	IUPAC name
$CH_3CH_2CH_2OH$	*n*- Propyl alcohol	1-Propanol
$(CH_3)_2CHCH_2OH$	Isobutyl alcohol	2-Methyl-1-propanol
$(CH_3)_2CHOH$	Isopropyl alcohol	2-Propanol

Methyl alcohol or methanol (CH_3OH) was formerly manufactured by the destructive distillation of wood. It is now produced synthetically from either coal or natural gas. Methyl alcohol is a volatile colourless liquid with a winelike odour. It is miscible with water in all proportions. It is poisonous and causes blindness. Methyl alcohol has been used as a solvent in some screen-printing inks.

Ethyl alcohol or ethanol is prepared largely by fermentation processes. The fermentation of glucose yields ethyl alcohol. Industrial ethyl alcohol is produced largely from the fermentation of molasses, the mother liquor left after the crystallization of cane sugar.

Ethyl alcohol is a colourless liquid having a sweet smell. It is commonly known as alcohol. It is miscible with water in all proportions. It is an intoxicant. Rectified spirit contains 96% alcohol. The alcohol sample which is completely free from water is called absolute alcohol. Mixture of lower alcohols containing methyl alcohol, ethyl alcohol, propyl alcohol and butyl alcohol may be obtained by the careful oxidation of natural gas. The individual alcohols may then be separated by fractionation.

Ethyl alcohol is used in both flexographic and gravure inks. Normal propyl alcohol and isopropyl alcohol are both used as solvents in the graphic arts for flexographic and screen-printing inks. Isopropyl alcohol is also used in lithographic dampening solutions.

Polyhydric alcohols   Alcohols containing more than one hydroxyl groups in their molecules are called polyhydric alcohols. Those alcohols having two hydroxyl groups per molecule are known as glycols. The principal glycol of commercial significance is ethylene glycol ($HOCH_2CH_2OH$, ethane-1, 2-diol), which is prepared from ethylene.

Ethylene glycol is a colourless, hygroscopic, viscous liquid. It is miscible with water and ethanol in all proportions but immiscible with ether. It is sweet in taste but toxic in effect. It forms low-freezing mixture with water and hence used as an antifreeze in automobile radiators. It is also employed in some screen-printing inks and in the manufacture of polyester film base.

Polypropylene glycol [$CH_3CH(OH)CH_2OH$, propane-1, 2-diol) resembles ethylene glycol in its methods of formation and properties. Industrially it is produced from propylene. However, unlike ethylene glycol, it is non-toxic. Hence, it is used in inks for food packaging. It is also used as a coolant in refrigerators. It finds use in the manufacture of polyester films. It is used as a plasticizer for cellophane.

The alcohols containing three hydroxyl groups are known as trihydric alcohols. The most important trihydric alcohol is glycerol or glycerine [$HOCH_2CH(OH)CH_2OH$, Propane-1, 2, 3- triol]. Glycerol is manufactured by the hydrolysis of fats and oils. It is a colourless, odourless, viscous and hygroscopic liquid. It is sweet in taste and miscible with water and ethanol in all proportions. But it is insoluble in organic solvents like ether, benzene, chloroform, etc.

Glycerol is often used as one of the ingredients in the dampening solution for lithographic duplicators. It is used in the preparation of alkyl resins and for esterifying rosins.

Aldehydes

Organic compounds having functional group —CHO are known as aldehydes. They are represented by the general formula, R—CHO, where R is an alkyl group. If R is an aryl group, then it is an aromatic aldehyde. The simplest aldehyde is formaldehyde, HCHO. Primary alcohols on oxidation give aldehydes.

Their common names are derived from the acids they give on oxidation. In IUPAC system an aldehyde is derived from the corresponding alkane by replacing the terminal "e" by "al". The longest carbon chain containing the aldehyde group is taken as the parent chain.

Formula	Corresponding acid	Common name	IUPAC name
HCHO	HCOOH (Formic acid)	Formaldehyde	Methanal
CH_3CHO	CH_3COOH (Acetic acid)	Acetaldehyde	Ethanal
C_2H_5CHO	C_2H_5COOH (Propionic acid)	Propionaldehyde	Propanal

Excepting formaldehyde which is gas, lower aldehydes are colourless volatile liquids. Higher members are solids. Aldehydes usually have an unpleasant smell. Lower members of the aldehyde are freely miscible with water but the solubility decreases rapidly with a rise in molecular weight. However, they are freely soluble in organic solvents. The aldehydes are strongly polar compounds.

Formaldehyde is used in the manufacture of plastics like bakelite and in the manufacture of dyes (such as pararosaniline, indigo, etc.). Acetaldehyde is used in the manufacture of acetic acid, acetic anhydride and ethyl acetate. It is also used in the manufacture of paraldehyde used as hypnotic and metaldehyde used as a solid fuel.

Ketones

Ketones are compounds having the general formula, R—CO—R′, where R and R′ are alkyl or aryl groups. When the organic groups are similar, the compound is known as a simple ketone. For example, dimethyl ketone, commonly called acetone, CH_3COCH_3. When the organic groups are different, the compound is known as a mixed ketone. For example, methyl ethyl ketone usually referred to as MEK, $CH_3COC_2H_5$.

Ketones are generally prepared by the oxidation of secondary alcohols. For example, acetone is obtained by the oxidation of isopropyl alcohol and methyl ethyl ketone is prepared by the oxidation of 2-butanol.

Ketones, unlike aldehydes, are not easily oxidized. Both aldehydes and ketones contain carbonyl group and therefore they show similarities in their properties.

Acetone is used extensively as a solvent for varnishes, lacquers, cellulose nitrate, cellulose acetate, etc. It is a good solvent for certain resins and is used in some screen-printing inks. MEK is a good solvent and is used in some flexographic, gravure and screen-printing inks.

Organic Acids

Organic acids are those compounds containing one or more carboxyl group, —COOH, which is a combination of carbonyl group and hydroxyl group. The acids are known as mono-, di-, tri- or polycarboxylic acids according to the number of carboxyl groups present in the molecule. They may be saturated or unsaturated, according to the nature of organic group attached to the carboxyl group.

The aliphatic saturated monocarboxylic acids form a homologous series, having the general formula, $C_nH_{2n+1}COOH$. The first member of this series is formic acid, $HCOOH$.

The next member is acetic acid, CH_3COOH. As many of these acids are constituents of fat, they are generally called as fatty acids.

According to IUPAC system the acids are named after the corresponding alkane by replacing the ending 'e' by "oic acid".

Formula	Common name	IUPAC name
$HCOOH$	Formic acid	Methanoic acid
CH_3COOH	Acetic acid	Ethanoic acid
C_2H_5COOH	Propionic acid	Propanoic acid

Organic acids are prepared by the oxidation of aldehydes. For example, formic acid is obtained by the oxidation of formaldehyde. They can also be obtained by the hydrolysis of esters. For instance, acetic acid is prepared by the hydrolysis of ethyl acetate ($CH_3COOC_2H_5$).

The first three members of fatty acids are colourless liquids having a pungent smell. They are soluble in water. The higher members are waxy solids, insoluble in water.

Formic acid is used in dyeing, tanning, coagulating rubber and in electroplating. Acetic acid is used in the preparation of pigments like white lead, copper acetate, etc. It is primarily used in the manufacture of cellulose acetate and vinyl acetate that are greatly used in the plastic industry.

The dicarboxylic acids contain two carboxyl groups.

Formula	Common name	IUPAC name
$(COOH)_2$	Oxalic acid	Ethanedioic acid
$CH_2(COOH)_2$	Malonic acid	Propanedioic acid
$(CH_2)_2(COOH)_2$	Succinic acid	Butanedioic acid

Hydroxy acids are those carboxylic acids, which contain both hydroxyl and carboxyl groups. They may be mono- or polybasic and may contain one or more hydroxyl groups. A number of the hydroxy acids have special names.

Hydroxy acids	Special name
$CH_2(OH)COOH$	Glycolic acid
$CH_3CH(OH)COOH$	Lactic acid
$HOOC.CH(OH)CH_2COOH$	Malic acid
$HOOC.CH(OH)CH(OH)COOH$	Tartaric acid
$HOOC.CH_2C(OH)COOH.CH_2COOH$	Citric acid

Esters

Esters are compounds formed by the reaction of acids and alcohols, similar to the reactions of acids and bases to form salts in inorganic chemistry. Such a reaction is called esterification. For example, ethyl acetate is prepared by the reaction of acetic acid with ethyl alcohol.

$$CH_3COOH + HOC_2H_5 \longrightarrow CH_3COOC_2H_5 + H_2O$$
Acetic acid   Ethyl alcohol   Ethyl acetate

The lower esters are colourless liquids having a fruity odour. The higher members are solids. The hydrolysis of an ester by an alkali is usually known as saponification.

They are generally named by first naming the alkyl group followed by the name of the acid and changing the "ic acid" by "ate". They have the general formula RCOOR′, where R and R′ may be same or different.

Formula	Common name	IUPAC name
$HCOOCH_3$	Methyl formate	Methyl methanoate
$CH_3COOC_2H_5$	Ethyl acetate	Ethyl ethanoate
$C_2H_5COOCH_3$	Methyl propionate	Methyl propanoate

Ethyl acetate is used in gravure, flexographic and screen-printing inks, and in the making of lacquers. Normal-propylacetate ($CH_3COOC_3H_7$) is used in gravure and flexographic printing inks. Isopropyl acetate [$CH_3COO.CH(CH_3)_2$] is used in gravure and screen-printing inks. Normal-butylacetate ($CH_3COOC_4H_9$) is used in screen-printing inks.

Ethers

They are the compounds having the general formula R—O—R′, where R and R′ are alkyl or aryl groups. When the R and R′ groups are similar, the compound is known as simple ether, e.g. diethyl ether, $C_2H_5OC_2H_5$. When the groups are not similar, the compound is called mixed ether, e.g. methyl phenyl ether or anisole, $CH_3OC_6H_5$.

Ethers are formed by treatment of alcohols with strong dehydrating agents (such as concentrated sulphuric acid). In the reaction, one molecule of water is removed from two molecules of alcohol. The two fragments of the alcohol join to form ether. The alkyl groups are joined through an oxygen atom.

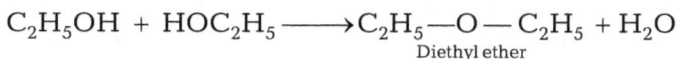

$$C_2H_5OH + HOC_2H_5 \longrightarrow C_2H_5-O-C_2H_5 + H_2O$$
Diethyl ether

In the common system, ethers are named by prefixing the names of alkyl groups attached to oxygen in alphabetical order before the word 'ether'. In IUPAC system ethers are considered as alkoxy derivatives of alkanes and named accordingly. In the case of asymmetrical ethers, the longest of the two alkyl groups is chosen as the parent hydrocarbon.

Formula	Common name	IUPAC name
CH_3OCH_3	Dimethyl ether	Methoxy methane
$CH_3OC_2H_5$	Ethyl methyl ether	Methoxy ethane
$C_2H_5OC_2H_5$	Diethyl ether	Ethoxy ethane

Ethers are mostly gases having a pleasant smell. They are highly volatile and sparingly soluble in water, but soluble in alcohol. Ethers are used widely as solvents. Diethyl ether is a very good solvent for fats, oils and resins.

The glycol ethers have both alcoholic and ether groups, and this combination gives them unusual solvent properties. The common example is ethylene glycol monoethyl ether, $HOCH_2CH_2OC_2H_5$. This glycol ether is used as a solvent in lacquers. It is used in some gravure, flexographic and screen-printing inks.

AROMATIC HYDROCARBONS

The aromatic organic compounds are all ring compounds or have cyclic groups of aromatic nature in their structure. The simplest aromatic compound is benzene.

Benzene

Benzene is the most important aromatic hydrocarbon. It is manufactured by the destructive distillation (heating in the absence of air) of coal. The structure of benzene is usually represented as

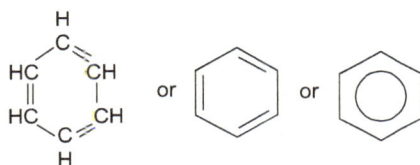

It is this ring structure with alternate single and double bonds that distinguishes an aromatic compound from an aliphatic compound.

Benzene is a colourless volatile liquid with a characteristic odour. It is soluble in water but insoluble in all organic solvents like ether, alcohol, etc. It is a good solvent for most organic substances (such as fats, oils, resins, etc.). It is an important starting material for the synthesis of dyes and plastics.

A wide variety of polycyclic aromatic hydrocarbons are known. They may have two or more benzene ring in the fused form. Some important examples are naphthalene, anthracene and phenanthrene. Naphthalene ($C_{10}H_8$) is a white crystalline compound derived from coal tar. Anthracene and phenanthrene are isomers ($C_{14}H_{10}$).

Naphthalene is used as solvents for varnishes and lacquers. Naphthalene and anthracene are widely used in the manufacture of dyestuff.

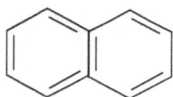
Naphthalene

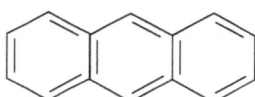
Anthracene

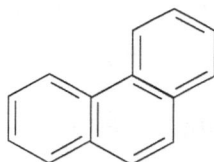
Phenanthrene

Benzene Derivatives

Benzene is the parent aromatic hydrocarbon. The homologues of benzene are formed by the replacement of one or more hydrogen atoms in the ring (nucleus) by alkyl groups. Thus if one hydrogen atom in the nucleus is replaced by a methyl group ($-CH_3$), methyl benzene or toluene, $C_6H_5-CH_3$ is formed. The monovalent group C_6H_5- is called phenyl group which is a typical aryl group. The alkyl groups like $-CH_3$ that replaces the hydrogen atom of the nucleus is called the side chain. Other benzene derivatives are similarly formed by the replacement of one or more hydrogen atoms of the nucleus by any functional group like $-Cl$, $-OH$, $-NH_2$, $-CHO$, $-COOH$, $-SO_3H$, etc. In any case benzene gives only one mono-substituted product since all the six hydrogen atoms in the regular hexagon ring are equivalent.

Some of the benzene derivatives are discussed below.

Toluene (phenyl methane or methyl benzene)   Toluene ($C_6H_5CH_3$) is the simplest alkyl derivative of benzene. It is

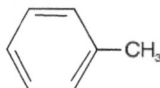
Toulene

obtained commercially from light petroleum fractions by catalytic cracking. It is a colourless liquid having a characteristic aromatic smell. It is insoluble in water but soluble in organic solvents. It is used as a solvent for paints, gums, lacquers, resins, etc.

Xylenes The xylenes are formed by the replacement of two hydrogen atoms in the benzene ring by methyl group (—CH₃). They are represented by the general formula $C_6H_4(CH_3)_2$. Depending on the relative positions of the two methyl groups on the ring, three kinds of xylenes are possible.

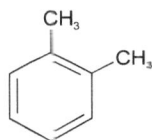

| *o*-Xylene | *m*-Xylene | *p*-Xylene |
| (1,2-dimethyl benzene) | (1,3-dimethyl benzene) | (1,4-dimethyl benzene) |

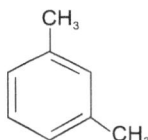

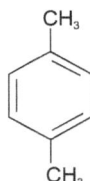

Ortho-, meta-, and para-xylene are abbreviated as *o*-, *m*- and *p*-xylene respectively. The commercial xylene (called xylol) consists of a mixture of these three isomers. It is used as a solvent in some gravure and screen-printing inks.

Phenol It is a monohydric phenol, which is also known as carbolic acid (carbo = coal, oleum = oil). Phenol is commercially obtained from the middle oil fraction of coal tar distillate.

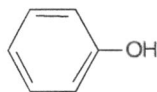

Phenol

Phenol is a colourless, crystalline solid which turns reddish brown on exposure to air and light. It has the smell of carbolic soap and a low melting point (42°C). It is sparingly soluble in water but readily soluble in many organic solvents. It causes burns when it comes into contact with skin.

Phenol is extensively used in the manufacture of synthetic resins (like bakelite) and certain azo dyes. It is also used as an ink preservative.

Polyhydric phenols   Three isomeric dihydric phenols are known. They are catechol, resorcinol and hydroquinone.

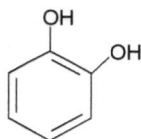   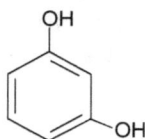   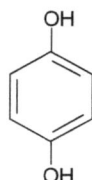

Catechol	Resorcinol	Hydroquinone
(Benzene 1,2-diol)	(Benzene 1,3-diol)	(Benzene 1,4-diol)

These are used in the manufacture of some dyes. Catechol and hydroquinone are used as photographic developing agents. Pyrogallol is a trihydric phenol. It is used as a photographic developing agent.

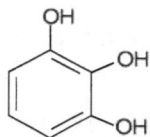

Pyrogallol
(1,2,3-trihydroxybenzene)

Aniline   It is a primary aromatic amine. It is obtained commercially by reducing nitrobenzene with iron and hydrochloric acid.

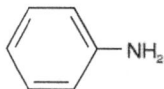

Aniline

When fresh, aniline is a colourless oily liquid. It is toxic and has a characteristic unpleasant odour. It is slightly soluble in water but soluble in alcohol, ether and benzene. Its colour becomes brown on standing due to oxidation.

Aniline is used in the preparation of diazonium compounds which are used in making azo dyes.

Toluidene It is also a primary aromatic amine. It has three isomeric forms, i.e. ortho-, meta-, and para-toluidene.

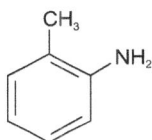

o-Toluidene	m-Toluidene	p-Toluidene
(o-Amino toluene)	(m-Amino toluene)	(p-Amino toluene)

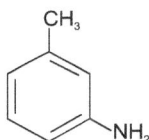

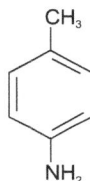

Benzoic acid Benzoic acid (C_6H_5COOH) is the simplest aromatic acid. It was first obtained from the gum, benzoin, and hence the name.

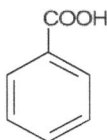

Benzoic acid

It is manufactured from toluene or xylene by oxidation. It is a white crystalline solid, sparingly soluble in cold water but fairly soluble in hot water, alcohol and ether. It is used as a food preservative in the form of its sodium salt.

FATS AND OILS

Fats and oils are the esters of glycerol with higher fatty acids and are known as glycerides. The glycerides of fatty acids that are liquid at ordinary temperatures are called oils and those that are solids are called fats. Chemically they are quite similar. The glycerides constituting fats contains a large proportion of saturated fatty acids such as palmitic acid ($C_{15}H_{31}COOH$) and stearic acid ($C_{17}H_{35}COOH$). On the other hand, glycerides constituting oils contain a large proportion of unsaturated acids such as oleic acid ($C_{17}H_{33}COOH$) and linolenic acid ($C_{17}H_{29}COOH$). The glycerides containing only one type of acid residue are called simple glycerides. Usually the glycerides containing more than one type of acid residue are known as mixed glycerides.

The glycerides are named according to the acid residues present in them by replacing the terminal 'ic acid' by 'in'. For example, stearin and palmitin are the glycerides of stearic and palmitic acid respectively. Similarly, olein and linolein are the glycerides of oleic acid and linoleic acid respectively.

Fatty acid	Fat/Oil molecule
Palmitic acid ($C_{15}H_{31}COOH$)	Palmitin ($C_{15}H_{31}COO)_3C_3H_5$
Stearic acid ($C_{17}H_{35}COOH$)	Stearin ($C_{17}H_{35}COO)_3C_3H_5$
Oleic acid ($C_{17}H_{33}COOH$)	Olein ($C_{17}H_{33}COO)_3C_3H_5$
Linoleic acid ($C_{17}H_{31}COOH$)	Linolein ($C_{17}H_{31}COO)_3C_3H_5$

Fats and oils are extracted from animals and plants. The following are some of the common oils and fats.

Vegetable oil	Animal oil	Vegetable fat	Animal fat
Linseed oil	Whale oil	Coconut oil	Lard
Soybean oil	Cod-liver oil	Vegetable ghee	Butter
Cotton seed oil	Fish oil	Palm oil	Beef tallow

Physical Properties

Fats and oils are colourless liquids or solids, which are lighter than water. They are insoluble in water, but soluble in organic solvents such as ether, chloroform, benzene, etc. They readily form emulsions when agitated with water in the presence of soap, gelatin or other emulsifiers.

Chemical Properties

Drying   When oils containing glycerides of unsaturated acids (such as linoleic and linolenic acid) are exposed to air, they undergo oxidation and polymerization to form a hard transparent coating. The process is known as drying and the oils are known as drying oils. Linseed oil, which is rich in linolenic acid, is a common drying oil used in varnishes and paints. The drying oil varnishes are used in printing inks.

Hydrolysis   Fats and oils can be hydrolysed to glycerol and the fatty acids by dilute acids, alkalis or superheated steam. The hydrolysis of oils and fats by alkalis is known as saponification.

Hydrogenation   Oils contain more of unsaturated glycerides than fats. When hydrogen is passed through them under pressure and in the presence of suitable catalyst, usually finely divided nickel, the unsaturated glycerides are converted in to saturated glycerides. Thus, oil is converted into solid fat. The process is known as hardening of oil.

Rancidification   When fats and oils are left exposed to moist air, they undergo decomposition and develop unpleasant smell. The process is known as rancidification. During rancidification, the chemical changes like oxidation and hydrolysis of fats and oils takes place.

Applications

They are used in the manufacture of soaps, detergents, glycerine, etc. Printing ink is made by grinding carbon black with oil containing a drier. Drying oils (like linseed oil, soybean oil, etc.) are extensively used in the manufacture of lacquers, varnishes and paints.

AMINO ACIDS

Amino acids are the amino-substituted acids containing both amino ($—NH_2$) and carboxyl ($—COOH$) groups. The simple amino acids have only one amino group and one carboxyl group in their molecules. These are classified as α-, β-, γ-, δ-, etc. amino acids depending on the relative position of amino and carboxyl group.

Class	Formula	IUPAC name (Common name)
α-Amino acid	H_2NCH_2COOH	Amino acetic acid (Glycine)
β-Amino acid	$H_2NCH_2CH_2COOH$	β-Aminopropionic acid
γ-Amino acid	$H_2NCH_2CH_2CH_2COOH$	γ-Aminobutyric acid

Among these α-amino acids are the most important and these are obtained by the hydrolysis of peptides or proteins.

Amino acids can also have more than one amino and carboxyl groups. These are classified as neutral, basic and acidic amino acids.

Neutral amino acids   These amino acids contain equal number of amino and carboxyl groups, e.g. cysteine, bis-(2-amino-2-carboxymethyl disulphide) ($HOOC.CH(NH_2)CH_2.S.S.CH_2CH(NH_2).COOH$).

Basic amino acids   These amino acids contain more number of amino groups than carboxyl groups, e.g. lysine, α,ε-diamino caproic acid ($H_2NCH_2CH_2CH_2CH_2CH(NH_2)COOH$).

Acidic amino acids   They contain more number of carboxyl groups than amino groups, e.g. aspartic acid, α-amino succinic acid ($HOOCCH_2CH(NH_2)COOH$).

Amino acids are mainly obtained by the hydrolysis of proteins. On boiling with dilute acids, proteins give a mixture of amino acids. These amino acids are then separated. They are crystalline solids with fairly high melting point. They are soluble in polar solvents (like water) and insoluble in non-polar solvents (like alcohol, benzene, etc.).

PROTEINS

Proteins are complex nitrogenous organic compounds containing C, H, O, S and P. They are the polymers of amino acids bonded by peptide linkages (—CO—NH—). Hence they are known as polypeptides.

Amino acid $-H_2O$ Amino acid

Protein

Their molecular weight ranges from 5000 to many millions. Many food materials are rich in protein. Some common commercial protein products are casein, soybean, gelatin and glue. There are twenty-three different α-amino acids from which all proteins are formed. Any given protein may have four to twenty three different amino acids in their structure. Proteins differ from fats and carbohydrates in having N and S atoms in their molecules. Proteins are isolated from suitable animal or vegetable source material.

Based on the chemical composition they are classified into two types.

Simple Proteins

They are made up of only chains of amino acids unit joined by peptide linkages, e.g. egg albumin, tissue globulin, glutenin, etc.

Conjugated Proteins

These have a non-protein part in their molecule. This non-protein part is known as the prosthetic group. Some of the conjugated proteins with their prosthetic group are given below.

Glycoproteins   They contain a carbohydrate or its derivative as prosthetic group, e.g. mucin in saliva.

Phosphoproteins   These have a phosphoric acid residue as prosthetic group, e.g. casein in milk.

Chromoprotein   They contain a coloured prosthetic group (usually a pyrrole derivative). The prosthetic groups usually contain a metal such as Fe, Cu, Mg, etc., e.g. haemoglobin in red blood cells.

General Properties of Proteins

  i. Most proteins are colourless amorphous solids. They do not have a sharp melting point.

 ii. Most of the proteins are insoluble in alcohol and water, but they are soluble in dilute acids and alkalis.

iii. Proteins form colloidal dispersion in water. They can readily pass through a filter paper but not through a membrane.

 iv. Proteins are easily precipitated by the action of heat, ethanol, concentrated inorganic acids, picric acid, UV and X-rays.

  v. Simple proteins can be hydrolysed with acids (HCl), alkalis (NaOH) or enzymes to give component amino acids. Conjugated proteins also yield the prosthetic group along with the amino acids.

 vi. Proteins on burning undergo oxidation, giving nitrogen, CO_2 and water.

Applications

Proteins like glue, casein and gelatin are industrially important materials Gelatin is widely used in the production of photographic emulsions that are employed in making photographic films. Gelatin is also used in food products. Casein is extensively used in the manufacture of paper, textile and adhesives. Glue is used as adhesive in wood industry and in sizing of paper. Proteins like wool and silk are widely used in textile industry. Proteins are also used in the production of various amino acids.

REVIEW QUESTIONS

1. Explain the general classification of organic compounds.
2. What do you understand by the following terms?
    i.   Aliphatic compound
    ii. Aromatic compound
    iii. Heterocyclic compound
    iv. Homologous series
    v.   Functional group
3. Classify the following compounds according to their functional groups:
    i. CH_3COCH_3
    ii. $C_2H_5OOCCH_3$
    iii. $C_2H_5OC_2H_5$
    iv. CH_3COOH
    v. CH_3CHO
    vi. C_2H_5OH
4. What are hydrocarbons? Discuss the types of hydrocarbons.
5. What is a homologous series? Explain the characteristic features of a homologous series.

6.   Write short notes on:
     i.   Alkanes
     ii. Alkenes
     iii. Alkynes

7.   Why are alkanes also known as paraffins?

8.   What are isomers? Write the structure of all possible isomers of a hydrocarbon with the molecular formula C_4H_{10}, C_4H_8 and C_5H_{12}.

9.   Write the structural formulae of the following.
     i.   Isopentane
     ii. 2, 2´-Dimethyl propane
     iii. 2-Pentene
     iv. 3-Methyl -1-butyne
     v.   Neopentane

10. Write a note on aromatic compounds.

11. Explain the following with suitable examples.
     i.   Alkyl halides
     ii. Ketones
     iii. Organic acids
     iv. Esters
     v.   Ethers
     vi. Aldehydes

12. What are alcohols? What do you understand by primary, secondary and tertiary alcohols?

13. What are polyhydric alcohols? Write the IUPAC name and structural formula of any two polyhydric alcohols.

14. Give the applications of the following in printing industry.
     i.   Glycol ether
     ii. Ethyl acetate
     iii. Methyl ethyl ketone
     iv. Glycerol
     v.   Ethylene glycol

15. Write the IUPAC name and structural formulae of the following.

    i. *p*-Xylene

    ii. Resorcinol

    iii. Pyrogallol

    iv. *m*-Toluidene

16. Write the name and structural formulae of all possible isomers of the following

    i. Xylene

    ii. Dihydric phenol

    iii. Toluidene

17. Write a note on polycyclic aromatic hydrocarbons.

18. What are fats and oils? Distinguish between fats and oils.

19. Explain the following properties of oils.

    i. Rancidification

    ii. Hydrolysis

    iii. Drying

    iv. Hydrogenation

20. What is meant by hardening of oils?

21. Give the applications of oils and fats.

22. What are amino acids? How are they related to proteins?

23. What are proteins? How are they classified?

24. Why are proteins also known as polypeptides?

25. Discuss the general properties of proteins.

26. Write a note on applications of proteins.

ADDITIONAL READING

Finar, I. L. (2004). *Organic Chemistry*, 6th edn. Vol. I & II. Pearson Education Ltd., Singapore.

Jones, Jr. M. (1997). *Organic Chemistry*. W.W. Norton & Co., London.

Morrison, R.T. and Boyd, R.N. (2004). *Organic Chemistry*. Pearson Education.

4   HIGH POLYMERS

INTRODUCTION

Polymers are materials which play a vital role in our everyday life and in industries as well. They provide the basic needs of our life such as food, clothings and shelters. Man's motivation to survive has made him to use the natural polymers like wood, proteins, starch, cotton, cellulose, natural rubber and leather all these years.

Polymers are so much a part of printing that it would be impossible to discuss the composition and properties of printing materials without considering the chemical nature of these giant molecules. Polymeric materials are widely used in the production of printing plates and rollers, blankets, photographic film base, and in packaging. They are also used in the formulation of printing inks, adhesives, etc.

The art of polymerization is of recent origin. The development of new techniques helped to produce new types of synthetic polymers with varied properties like high strength, good flexibility, and resistance to heat, chemicals and solvents. These can be regulated to suit the requirements. Therefore, polymers have become important engineering materials. The first modified natural polymer, cellulose nitrate, was commercially produced around 1860 and the first synthetic polymer phenol–formaldehyde was made around 1910. The major systematic development of the present-day polymer science and technology has taken place since 1920 and this leads to the production of a large number of polymers of multifarious utility.

FUNDAMENTALS OF POLYMERS

Polymers (Greek: *poly-mer* means "many parts") are giant molecules formed by the repeated linking of a large number of simple molecules.

The simple molecules of low molecular weight which combine with each other to form a polymer molecule are called **monomers**. In other words, simple monomers are the building blocks of complex polymers.

Generally polymers are formed by sequential addition of many monomers in a regular manner by the process known as **polymerization**. For example, polyethylene is a polymer formed by the repeated linking of a large number of ethylene molecules. The total number of repeating monomer units (n) present in a polymer molecule is known as the **degree of polymerization**.

The number of bonding sites or functional groups present in a monomer molecule is called as the **functionality** of the monomer. A monomer can undergo polymerization, if and only if, it is bifunctional. Thus monomers may be bifunctional or polyfunctional depending on the number of bonding sites present. For example, all vinyl monomers ($H_2C{=}CHX$, where, X is H, CH_3, C_6H_5, Cl, etc.) contain two bonding sites (one double bond is equivalent to two bonding sites) and hence are bifunctional. The structure of a polymer depends on the functionality of its monomers. Thus polyfunctional monomers usually form cross-linked polymers.

CLASSIFICATION OF POLYMERS

Polymers may be classified in different ways.

Classification Based on Occurrence

On the basis of occurrence, polymers are classified as natural and synthetic polymers.

Natural polymers Polymers which are obtained from natural sources such as vegetables and animals are called natural polymers.

Examples Cotton, silk, wool, cellulose, starch, proteins, natural rubber.

Synthetic polymers Polymers that are obtained by synthetic methods from low-molecular-weight monomers are known as synthetic polymers.

Examples Polyethylene, nylons, terylene, polyvinyl chloride polyurethane, phenol–formaldehyde resin (Bakelite).

Classification Based on Structure

Based on their structure, polymers are classified as linear, branched-chain and cross-linked polymer.

Linear polymers In the linear polymer, each monomer is linearly linked to only two other monomers. In the case of a bifunctional monomer, two reactive groups attach side by side to each other forming a linear or straight-chain polymer. If 'M' represents a monomer, the typical structure of a linear polymer is represented as

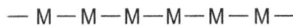

$$- M - M - M - M - M - M -$$

Examples High-density polyethylene (HDPE), polystyrene, polyvinyl chloride, polytetrafluoroethylene (Teflon).

Branched-chain polymers During the chain growth, the linear polymeric chains may branch out and grow sideways along with the parent chain. If there is any branching on the polymer chain, it is referred to as branched-chain polymer. The typical structure of a branched-chain polymer may be represented as

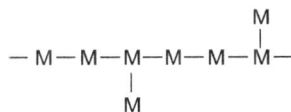

$$\begin{array}{c} M \\ | \\ - M - M - M - M - M - M - \\ | \\ M \end{array}$$

Examples   Low-density polyethylene (LDPE).

A branched-chain polymer is also formed, when a trifunctional monomer is polymerized in the presence of a small amount of a bifunctional monomer.

Examples   Copolymer of vinyl chloride and styrene.

Cross-linked polymers In the polymerization of polyfunctional monomers, branching and cross-linking usually occurs. This results in the formation of a polymer with three-dimensional network type structures. Such polymers are known as cross-linked polymers. The typical structure of a cross-linked polymer is represented as

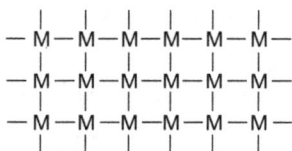

$$
\begin{array}{cccccc}
\mid & \mid & \mid & \mid & \mid & \mid \\
-\text{M}-\text{M}-\text{M}-\text{M}-\text{M}-\text{M}- \\
\mid & \mid & \mid & \mid & \mid & \mid \\
-\text{M}-\text{M}-\text{M}-\text{M}-\text{M}-\text{M}- \\
\mid & \mid & \mid & \mid & \mid & \mid \\
-\text{M}-\text{M}-\text{M}-\text{M}-\text{M}-\text{M}- \\
\mid & \mid & \mid & \mid & \mid & \mid
\end{array}
$$

Examples   Phenol–formaldehyde, urea–formaldehyde, epoxy resins.

Classification Based on the Method of Polymerization

Based on the method of polymerization, polymers are classified as addition polymers and condensation polymers.

Addition polymers The polymers which are formed by addition polymerization reaction are called addition polymers. Addition polymers made up of only one type of repeating units (monomers) are known as homopolymers while those polymers formed from more than one type of monomers are referred to as copolymers.

Examples   *Homopolymers*   Polyethylene, polystyrene, polypropylene, polyvinyl chloride, polyvinyl acetate.

Copolymers   Buna-N, Buna-S, butyl rubber.

Condensation polymers The polymers obtained by condensation polymerization reaction are known as condensation polymers.

Examples Bakelite, terylene, nylon 6,6, epoxy resins.

Classification Based on Thermal Behaviour

Based on their thermal behaviour, they are classified as thermoplastics and thermosettings.

Thermoplastics Polymers which soften on heating and harden on cooling are called thermoplastics. Repeated heating and cooling do not alter the chemical nature of these polymers, because the changes involved are purely of physical nature.

Examples Polyethylene, polystyrene, polyvinyl chloride, nylons.

Thermosetting polymers Polymers which soften initially and undergo chemical changes on heating, and become rigid, infusible and insoluble mass on cooling are called thermosetting polymers. They set in the first heating itself and hence cannot be remoulded.

Examples Phenol–formaldehyde, urea–formaldehyde, epoxy resins.

Classification Based on Properties and Uses

Based on their properties and uses, polymers are classified as plastics, fibres, elastomers and resins.

Plastics Polymers which can be moulded into the desired shape by application of heat and pressure are termed as plastics.

Examples Polyethylene, polystyrene, polyvinyl chloride.

Fibres Polymers which can be used as fibres belong to this type. Fibres have a long, thin, and thread-like molecules. These long-chain polymers are lined up and held together through stronger intermolecular forces (hydrogen bonding).

They may be obtained either by natural sources or synthetic methods.

Examples   *Natural fibres*   Cotton, wool, silk, jute.

Synthetic fibres Nylons, terylene.

Elastomers   Polymers which show a high degree of elasticity are called elastomers. They undergo elongation on stretching and regain their original shape on releasing the force. The elasticity of these polymers is due to weak intermolecular forces between polymer chains which undergo uncoiling on application of force.

Examples   Natural rubber and synthetic rubbers like Buna-S, Buna-N, butyl rubber, silicone rubber.

Resins   They are the low-molecular-mass polymers that exist in semi-solid or liquid form and are commonly used in adhesives, paints, varnishes, etc.

Examples   *Natural resins*   Shellac, copal.

Synthetic resins Phenol–formaldehyde, urea–formaldehyde, and epoxy resins.

TYPES OF POLYMERIZATION

The rapid growth of the plastic industry in recent years has been largely based on completely synthetic polymers (like polyethylene, polystyrene, nylons, etc.). The synthetic polymers are formed by two basic types of polymerization reactions.

1. Addition (chain) polymerization
2. Condensation (step) polymerization

Addition (Chain) Polymerization

In this polymerization reaction, monomers containing one or more double bonds undergo rapid self-addition to each other without the elimination of any by-product. The reaction usually occurs in the presence of a small amount of catalyst known as initiators. In general, polymers of olefins, vinyl derivatives, dienes, etc. are obtained by addition polymerization reaction, for example,

$$n \; H_2C{=}CH_2 \xrightarrow[\text{Initiator}]{\text{Heat/light}} \left[H_2C - CH_2 \right]_n$$

Ethylene   Polyethylene

Addition polymerization produces linear polymers. The molecular weight of the polymer obtained is an integral multiple of that of the monomer.

This polymerization reaction takes place by free-radical, ionic or coordination mechanism. Each of these mechanisms involves three steps: initiation, propagation and termination. Thermally unstable peroxides such as hydrogen peroxide, benzoyl peroxide, etc. are used as initiators in free-radical polymerization. In ionic polymerization $NaNH_2$, BF_3, $AlCl_3$, etc. are used as initiators.

The initiators are able to react with the double bond in a monomer to produce free radicals or ions. These free radicals or ions, then add to the double bond of another monomer unit to form a new free radical or ion which in turn adds to another monomer unit leading to continuous growth of the polymer chain by process of chain propagation. Finally, the chain terminates if an ionic polymeric chain continues with an ion of opposite charge, and in the case of free radical polymeric chain, or combination with another free radical.

In coordination polymerization, Ziegler–Natta catalyst is used. The Ziegler–Natta catalyst consists of an organometallic compound (usually trialkyl aluminum) and a transition metal halide (usually $TiCl_4$). The polymerization reaction usually occurs at room temperature and at lower pressure. It forms a stereoselective polymer which is linear and mechanically stronger.

The addition polymerization of two or more different monomers is known as copolymerization. The high polymers obtained by this polymerization reaction are called copolymers. For example, 1, 3-butadiene and styrene copolymerize to produce Buna-S.

Some important addition polymers are the following.

Polyethylene   Polyethylene is obtained by the addition polymerization of ethylene.

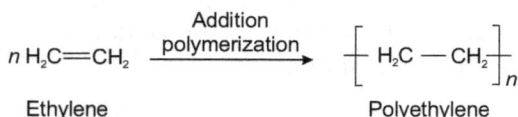

$$n\ H_2C{=}CH_2 \xrightarrow[\text{polymerization}]{\text{Addition}} {-}\!\!\left[H_2C - CH_2 \right]_n$$

Ethylene   Polyethylene

Two commercial grades of polyethylene are obtained under different conditions. When ethylene is polymerized under high pressures (about 1500 atm.) and under high temperatures (150–250°C) in the presence of traces of oxygen as initiator, low-density polyethylene (LDPE) is obtained. LDPE is a branched-chain polymer with low density and softening point (110–120°C).

If ethylene is polymerized under low pressures (6–7 atm.) and at low temperatures (60–70°C) using Ziegler–Natta catalyst (a mixture of $Al(C_2H_5)_3$ and $TiCl_4$ in n-heptane), high density polyethylene (HDPE) is formed. HDPE is a linear polymer with high density and softening point (140–150°C). It is more crystalline and chemically inert than LDPE.

Polyethylene films are mainly used for packaging, waterproofing, coating, and lamination. HDPE can be steam sterilized because its softening point is well above the temperature of boiling water. Hence, it is used for packaging of foodstuff.

Polypropylene   It is manufactured by the polymerization of propylene using Ziegler–Natta as a catalyst in n-heptane.

$$n \; H_2C=\!\!=\!CH \quad \xrightarrow[\text{catalyst}]{\text{Ziegler–Natta}} \quad \left[H_2C - CH \right]_n$$
$$\qquad\qquad CH_3 \qquad\qquad\qquad\qquad\qquad CH_3$$

Propylene   Polypropylene

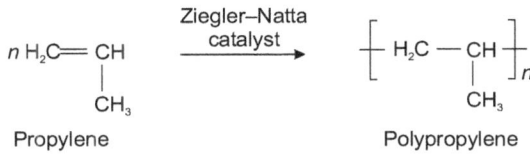

It has high crystallinity, stiffness, tensile strength, and softening point (160–170°C). It is chemically inert and a good electrical insulator.

Polypropylene films and sheets are used as packaging materials in food, fertilizer and cement industry. Coated and co-extruded forms of polypropylene are widely used in food packaging. Polypropylene coated with polyvinylidene chloride exhibits high gas and moisture barrier properties. It is suitable for deep-freeze applications. Acrylic-coated polypropylene provides heat-sealable films. A co-extruded polypropylene film consists of a central core of polypropylene with its outer surface modified by heat-fusing with polyethylene. These heat-sealable films find applications in packaging of foodstuff such as crisps, snack foods and chocolate products. Polypropylene is one of the plastics successfully used in the injection moulding of duplicate printing plates. Thick, rigid films of polypropylene are used as backing layers for printing plates.

Polystyrene Polystyrene is manufactured by the bulk polymerization of styrene using benzoyl peroxide as initiator.

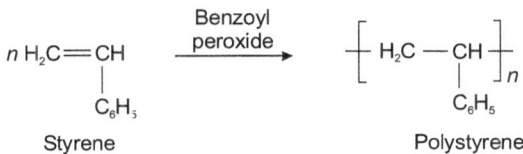

$$n \; H_2C=\!\!=\!CH \quad \xrightarrow[\text{peroxide}]{\text{Benzoyl}} \quad \left[H_2C - CH \right]_n$$
$$\qquad\qquad C_6H_5 \qquad\qquad\qquad\qquad\qquad C_6H_5$$

Styrene   Polystyrene

Polymerization can be carried out through free radical, ionic or coordination mechanism.

Polystyrene is a transparent and chemically inert thermoplastic. It has excellent optical properties.

Polystyrene films are widely used for packaging. It is also commonly used in sheets for printing backlit point-of-purchase displays. It is used in the manufacture of foamed products. These foamed products provide good heat insulation and hence are used in packaging.

Polyvinyl chloride (PVC)   It is an addition polymer of vinyl chloride and manufactured by emulsion polymerization method. An emulsion of vinyl chloride in water (soap as an emulsifier) is polymerized in the presence of peroxides such as hydrogen peroxide (H_2O_2).

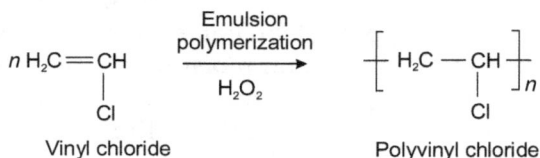

$$n \, H_2C{=}CH \xrightarrow[H_2O_2]{\text{Emulsion polymerization}} \left[H_2C - CH \right]_n$$

$$\quad\quad\quad | \quad\quad\quad\quad\quad\quad\quad\quad\quad\quad\quad |$$
$$\quad\quad\quad Cl \quad\quad\quad\quad\quad\quad\quad\quad\quad\quad Cl$$

Vinyl chloride   Polyvinyl chloride

It is a colourless, non-inflammable and chemically inert powder. It dissolves in chlorinated hydrocarbons. Due to the existence of high intermolecular attraction forces, commercial PVC is a hard and stiff amorphous plastic. Hence, it is difficult to process. It is therefore compounded with plasticizers (such as dioctyl phthalate, tricresyl phosphate, etc.). The product thus obtained is called plasticized PVC, which is more flexible.

PVC is used in making press rollers for letterpress and offset printing. Apart from its applications in packaging and plastic binding, PVC is widely used in duplicate plate making. It is also used in making vinyl tapes, table covers and vinyl floorings. It finds use in making inks which are employed for printing on vinyl substrates.

Polyvinyl acetate   It is obtained by the polymerization of vinyl acetate (prepared by treating acetylene with acetic acid in presence of $HgSO_4$ as catalyst) in presence of benzoyl peroxide or hydrogen peroxide as catalyst.

It is a colourless, transparent material, and is resistant to water, atmospheric oxygen and common chemicals. It has relatively a low softening point. It is fairly soluble in organic acids and organic solvents.

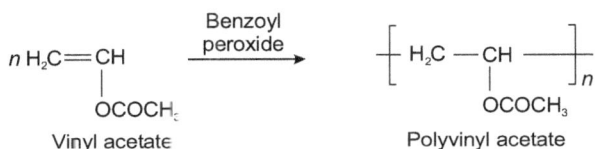

$$n\,H_2C=CH \quad \xrightarrow{\text{Benzoyl peroxide}} \quad \left[\!\!\begin{array}{c} H_2C-CH \\ | \\ OCOCH_3 \end{array}\!\!\right]_n$$

OCOCH₃ (Vinyl acetate) → Polyvinyl acetate

Vinyl acetate   Polyvinyl acetate

Polyvinyl acetate is used in the manufacture of polyvinyl alcohol, water-based emulsion paints and adhesives. It is also used in foil lacquers and heat-sealing lacquers for flexographic applications. It is used as a sizing material and as a "permanent starch".

Polyvinyl alcohol Polyvinyl alcohol is prepared by the polymerization reaction of polyvinyl acetate with an alcohol in presence of an acid or alkali as catalyst.

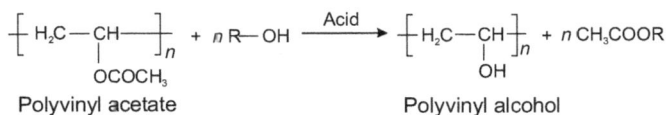

$$\left[\!\!\begin{array}{c} H_2C-CH \\ | \\ OCOCH_3 \end{array}\!\!\right]_n + n\,R-OH \quad \xrightarrow{\text{Acid}} \quad \left[\!\!\begin{array}{c} H_2C-CH \\ | \\ OH \end{array}\!\!\right]_n + n\,CH_3COOR$$

Polyvinyl acetate   Polyvinyl alcohol

Polyvinyl alcohol is soluble in water. It is widely used in adhesives, paper coatings, paper sizing and textile sizing. The completely hydrolysed grades of polyvinyl alcohol find use in quick-setting, water-resistant adhesives. The combination of fully hydrolysed polyvinyl alcohol and starch are used as quick-setting adhesives for paper converting. Partially hydrolysed grades have been developed for making tubular blown film for packaging bleaches, bath salts, insecticides and disinfectants. Polyvinyl alcohol sensitized with a dichromate salt is widely used in photoengraving. It is used for giving oxygen barrier coating on packaging materials. It is one of the constituents in the photoemulsions used in stencil making in screen-printing. It is used in some light-sensitive coatings for lithographic plates.

Condensation (Step) Polymerization

In condensation polymerization reaction, monomers containing two or more reactive functional groups undergo intermolecular condensation to produce high polymers. The condensation of monomers takes place with the continuous elimination of by-products such as H_2O, NH_3, HCl, C_2H_5OH, etc. Initially, bifunctional monomers react to form a dimer which still has two free functional groups, one at each end. This dimer is able to react with monomers at both of these ends to a tetramer which again has two free functional groups, one at each end. Thus, there is an increase in the chain length by stepwise growth resulting in a high polymer. Hence, this polymerization reaction is also known as step polymerization. Unlike addition polymerization, this polymerization reaction is a slow process. The condensation polymerization reactions are usually catalysed by acids and/or alkalis and produce linear or cross-linked polymers.

Typical examples of polymers obtained by condensation polymerization are polyamides (nylon 6, 6), polyesters (terylene), phenol–formaldehyde, urea–formaldehyde, epoxy resins and polyurethanes.

Some important condensation polymers are the following.

Polyesters   They are prepared by condensation polymerization of dicarboxylic acids with dihydroxy alcohols. For example, terylene (dacron) is a saturated polyester obtained by the condensation reaction of ethylene glycol with terephthalic acid in the presence of an acid catalyst and under high temperature.

It can also be obtained by condensation of ethylene glycol with diethyl or dimethyl terephthalate.

Polyesters have high tensile strength, and high crease- and wrinkle resistance. They are chemically inert towards minerals and organic acids, but are acted upon by alkalis. They are heat-resistant with a softening point of 230–240°C.

n HO—CH$_2$—CH$_2$—OH   + n HOOC —⟨◯⟩— COOH

Ethylene glycol   Terephthalic acid

H$^+$ | heat

$$\left[\!\!-\text{O—CH}_2\text{—CH}_2\text{—O—OC}\!-\!⟨◯⟩\!-\!\text{CO}-\!\!\right]_n \; + \; 2n\,\text{H}_2\text{O}$$

Terylene

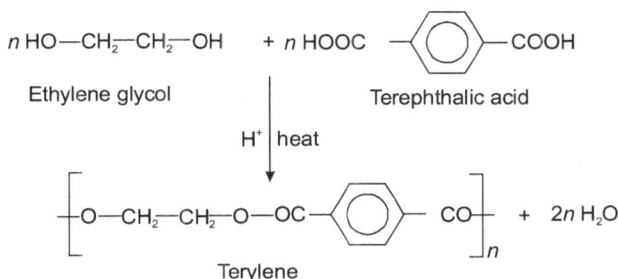

Polyester films are used in the manufacture of photographic films and magnetic recording tapes. Because of their chemical inertness, polyester films are extensively used in packaging the laminates, pharmaceuticals, and liquid sachets. They are used as the base materials for the letterpress printing plates. Other applications of polyester film in the printing industry are in letterpress cylinder dressing, as a drafting film in the preparation of originals from which printing plates or diazo copies are produced, in colour proofing systems and as a base for low-cost pre-sensitized lithoplates. Polyester films are also used as substrates in photostencil film for screen-printing in which polyester films are coated with a UV-reactive photopolymer. Polyester fibres are widely used in textile industry.

Polyamides Polyamides are the synthetic polymers with amide groups (—CO—NH—) in their structure. They are often referred to as nylons. The typical example is nylon 6,6. It is manufactured by the condensation of hexamethylene diamine with adipic acid at about 300°C and under pressure.

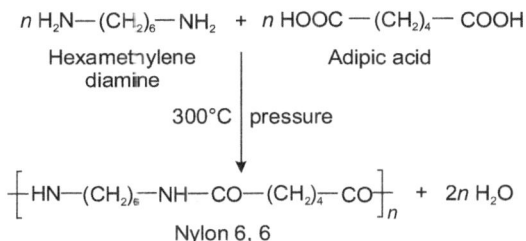

n H$_2$N—(CH$_2$)$_6$—NH$_2$ + n HOOC —(CH$_2$)$_4$— COOH

Hexamethylene   Adipic acid
diamine

300°C | pressure

$$\left[\!\!-\text{HN—(CH}_2)_6\text{—NH—CO—(CH}_2)_4\text{—CO}-\!\!\right]_n \; + \; 2n\,\text{H}_2\text{O}$$

Nylon 6, 6

Each monomer contains six carbon atoms and hence it is called nylon 6,6.

Nylon 6,6 is a strong, tough and elastic polymer. It has high thermal stability (softening point, 235°C), good abrasion resistance and solvent resistance. But it dissolves in phenol and formic acid.

Polyamide films are extensively used for packaging foodstuff and pharmaceuticals, because of their high gloss and fat resistance, and are heat-sealable in nature. Polyamides are widely used in flexographic and gravure inks for printing on flexible packaging because they provide high adhesion to corona-treated films coated with polyethylene, polystyrene, polypropylene and polyvinylidene chloride. Polyamides also find use in hot-melt adhesives and in aqueous adhesive dispersions. They are also used in overprint varnishes where they impart scratch resistance. They find applications in making printing plates both in photopolymer systems and in duplicate relief plates. Polyamides are used as coating materials for plastic films and paper where they not only provide grease and water-resistance barriers for packaging products but also enable them to be heat-sealed.

Phenolic resins   Phenolic resins form a very important group of thermosetting polymers. These are the condensation products of phenols with an aldehyde. The product obtained depends on the concentration and chemical nature of the reactants, the type of catalyst used, and the temperature.

Phenol–formaldehyde (Bakelite) is an important phenolic resin. It is manufactured by the condensation polymerization of phenol (C_6H_5OH) with formaldehyde (HCHO) in the presence of an acid or alkali as catalyst. The reaction involves the following steps.

The initial reaction results in the formation of mono-, di- and tri-methylol phenols.

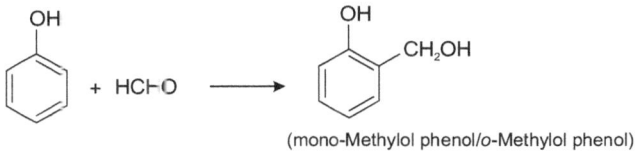

(mono-Methylol phenol/o-Methylol phenol)

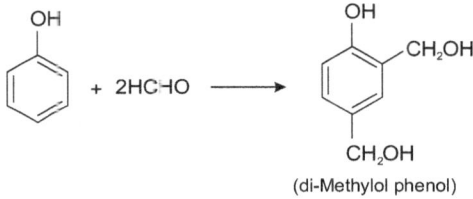

(di-Methylol phenol)

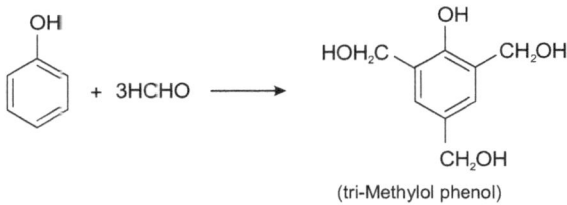

(tri-Methylol phenol)

The further reaction between methylol phenols and phenol takes place in two ways resulting in the formation of two types of resins

1. Condensation between hydroxyl group of methylol phenol and hydrogen of the benzene ring of phenol with the elimination of water.

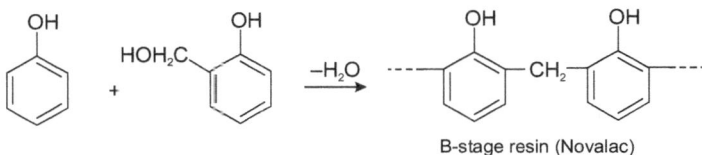

B-stage resin (Novalac)

When the ratio of phenol to formaldehyde is greater than 1 (i.e., phenol is taken in excess over formaldehyde) and in presence of an acid as a catalyst, B-stage resin (Novalac) is formed. It is characterized by methylene linkage.

2. Condensation between two methylol groups of methylol phenols with the elimination of water.

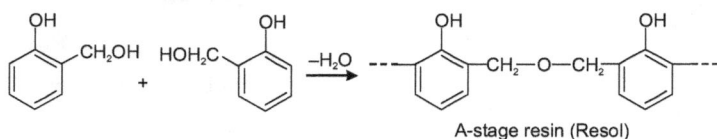

A-stage resin (Resol)

The resulting resin is called A-stage resin (Resol). This is formed when the phenol to formaldehyde ratio is less than 1 (i.e., formaldehyde is taken in excess over phenol) and in presence of an alkali as a catalyst. It is characterized by methylene–ether linkage.

These A-stage and B-stage resins during moulding process (in presence of a curing agent) undergo cross-linking to produce a hard, rigid, infusible solid called C-stage resin. This is commonly known by the trade name Bakelite.

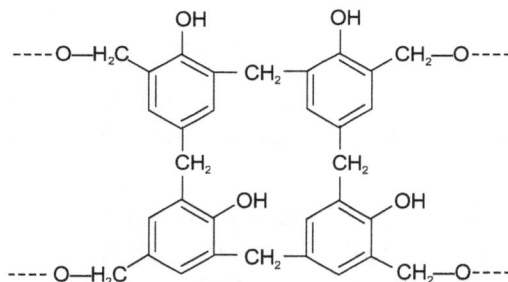

Bakelite

Bakelite is a rigid, hard, infusible, insoluble, water-resistant, and scratch-resistant thermosetting resin. It has good dimensional stability, thermal resistance and chemical resistance except for alkalis.

Phenol–formaldehyde resins are mainly used in making printing inks, varnishes, lacquers, paints and adhesives. They are used for impregnating paper, wood, fabrics, etc. They are widely used in the manufacture of laminated plastics. The varnish produced by the reaction of phenolic resins with tung oil provides good adhesion, fast drying and very tough gloss films. Rosin-modified phenolic resins are used in all types of inks because of their high gloss and good rub-resistance properties.

Alkyd (Glyptal) resins The condensation polymerization of polybasic acids (such as phthalic anhydride) with polyhydric alcohols (such as glycerol) on heating under high pressure and in presence of an acid catalyst produces alkyd resins. The reaction between phthalic anhydride and glycerol first forms a linear polymer which later on by cross-linking of chains gives rise to a three-dimensional network type structure. Thus by changing the acid and alcohol components, and conditions of the reaction, it is possible to get both thermosetting as well as thermoplastic type alkyd resins.

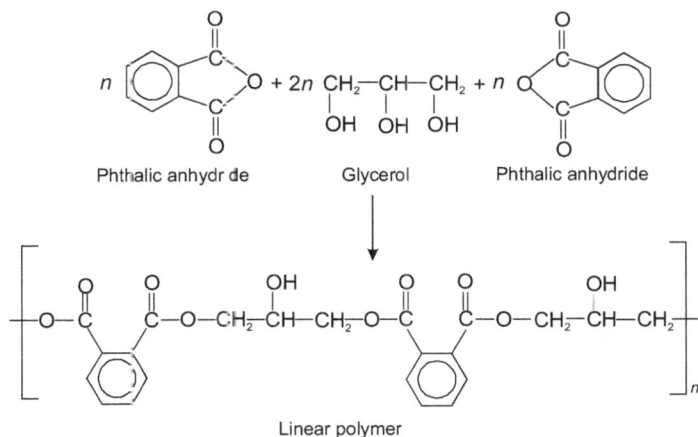

Alkyd resins can be modified either by reacting with drying or non-drying oils or by fatty acids. These modified resins dry rapidly and gives a tougher and more flexible film with superior gloss. They possess good pigment wetting characteristics.

They are widely used in making alkyd varnishes. These varnishes are employed in lithographic, screen- and letterpress printing. Drying-oil-modified alkyds are used as surface-coating materials. They also find use as resin plasticizers in the form of their lower polymer states. Alkyds modified with non-drying oils make excellent plasticizers for nitrocellulose in gravure inks. Alkyd resins are used mainly as cements, binder and in paints.

Epoxy resins   Epoxy resins are the condensation products of bisphenol A and epichlorohydrin formed in the presence of alkali (catalyst).

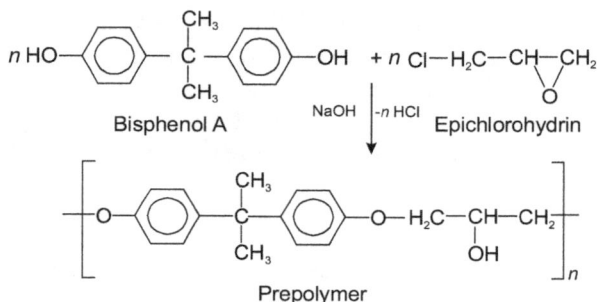

Prepolymer

These prepolymers are cross-linked (cured) at room temperature on mixing with diamines (such as *m*-phenylene diamine or hexamethylene diamine). The process is generally called cold curing.

Epoxy resins show resistance to water, acids, alkalis, various solvents and environmental degradation. The polar nature of the resin molecules provides excellent adhesion property to them. The cured resin shows high flexibility, toughness and heat resistance.

Epoxy resins are widely used as adhesives. They are used in the production of varnishes, paints, and glass-fibre-reinforced plastics. The good adhesion and water resistance properties of epoxy resins make them ideal for use as coating materials. They are used for coating on gravure cylinders. The resin is used in making epoxy inks that are employed when good adhesion and solvent resistance are required on substrates like glass, metals, ceramics and certain plastics.

Cold-cured epoxy resins are used in the plastic-backed "college electrotypes", which are thinner and more flexible than a conventional electrotype. They are also used as laminating materials.

Polyurethanes Polyurethanes are characterized by the presence of urethane linkages (—NH—CO—O—) in their molecular chains. In the industry, plastic-like polyurethanes are known by the trade name "Igamid" while fibre-like polyurethanes are called "Perlon".

Polyurethanes are obtained by the condensation polymerization of a diisocyanate with a polyhydroxy alcohol in the presence of tertiary amines (such as triethylene diamine as the catalyst).

Perlon-U is obtained by the condensation reaction of 1, 6-hexamethylene diisocyanate with 1, 4-butanediol. Other diisocyanates used are toluene 2, 4-diisocyanate, 4, 4-diisocyanato diphenyl methane, etc.

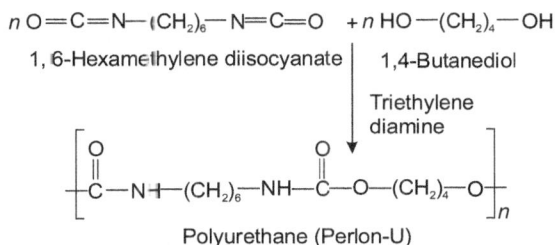

$$n\ O{=}C{=}N{-}(CH_2)_6{-}N{=}C{=}O \quad +n\ HO{-}(CH_2)_4{-}OH$$

1, 6-Hexamethylene diisocyanate | 1,4-Butanediol

Triethylene diamine

$$\left[\begin{array}{c} O \\ \| \\ C{-}N{+}(CH_2)_6{-}NH{-}\overset{O}{\overset{\|}{C}}{-}O{-}(CH_2)_4{-}O \end{array} \right]_n$$

Polyurethane (Perlon-U)

Polyurethanes have good resistance to oils, grease, and solvents. They have good elasticity and excellent resistance to abrasion. These are less stable than polyamides at higher temperatures.

Polyurethanes are mainly used in the production of flexible foams, adhesives and ink rollers. They are employed as surface coatings and packaging materials. They are used in making modified drying oil varnishes which are employed in sheet-fed lithographic inks.

Polycarbonates   They are polyesters of unstable carboxylic acids. They are prepared by condensation reaction of diphenyl carbonate and bisphenol A [2, 2-bis(4-hydroxy phenyl) propane].

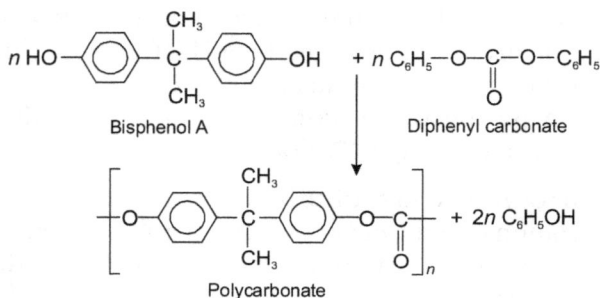

Bisphenol A   Diphenyl carbonate

Polycarbonate

They are transparent thermoplastics, which possess high impact and tensile strength over a wide range of temperature, good dimensional stability and stiffness. They also have a good resistance to heat and flame.

They are used in making sterilizable transport containers. They are presently used for making compact audio discs. They are used as base materials in the manufacture of photographic films and all types of optical discs.

PLASTICIZERS

Plasticizers are low-molecular-weight non-volatile substances (mostly liquids), which when added to a polymer improve its flexibility, processibility and hence utility. They substantially reduce the brittleness of many amorphous polymers because their addition even in small quantities markedly reduces the glass transition temperature (T_g)—the temperature at which a polymer changes from the glassy state to the rubbery state of the polymer. This effect is due to a reduction in cohesive forces of attraction between polymer chains. The plasticizer molecules penetrate into the polymer matrix and establish polar attractive forces between it and chain segments. These attractive forces reduce the cohesive forces between the polymer chains and increase the segmental mobility, thereby reducing the T_g value. The commonly used plasticizers are vegetable oils, mineral oils, stearic acid, wax, coal-tar oils, tricresyl phosphate, triphenyl phosphate, and dibutyl phthalate.

Some polymer materials such as polyvinyl chloride are stiff and brittle, and cannot be moulded below their decomposition temperature, so they must be softened. Plasticizers are used to soften them. They are usually used as one of the ingredients in the compounding of rubber. Plasticizers are incorporated in flexographic and gravure inks, and in most surface coatings drying by solvent evaporation. The presence of a plasticizer provides a flexible film and also improves the adhesion of the film to the substrate. They are also employed in the formulation of rubber rejuvenators (a deglazer material used to remove glaze from rubber rollers).

Plastisols are suspensions of vinyl resins in plasticizer. They remain stable at cold, but when heated the resin dissolves forming a film of plasticized vinyl resin. An ink formulation added with plastisol gives a tough vinyl film on heating, which provides good properties for the dried ink. Plastisol inks are used in screen-printing.

ELASTOMERS

Elastomers are polymers having rubber-like properties, i.e., when stretched they elongate to many times their length and the original length is restored on releasing the force. This property of elastomers is called elasticity. Natural and synthetic rubbers belong to this class of polymers.

An elastomer is a linear polymer with highly coiled structure. A very weak type of intermolecular forces of attraction is involved between the chains. The unstretched rubber is amorphous. On stretching, the polymer chains get partially aligned with respect to another, thereby causing crystallization. This enhances the attraction forces between different chains thereby causing stiffening of the material. On releasing the deforming stress, the chains get reverted back to their original coiled state and the material again becomes amorphous.

Natural Rubber

Natural rubber is a polyisoprene having molecular formula, $(C_5H_8)_n$. The degree of polymerization of natural rubber is about 5000. Polyisoprene exists in two isomeric structures, *cis-* and *trans-* forms. Natural rubber is soft and has *cis*-configuration. Gutta percha is a hard rubber and exists in *trans*-configuration.

Natural rubber is found in different species of rubber trees. Among these, *Hevea brasiliensis* is the most important one and the rubber obtained from this is often called "Hevea rubber". It is obtained in the form of milky fluid known as latex from incisions in the bark of a rubber tree. Latex contains 30–36% rubber hydrocarbon, 2–4% proteins, 1–2% resins and the rest is water.

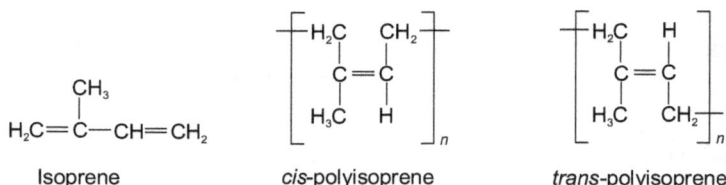

Isoprene   *cis*-polyisoprene   *trans*-polyisoprene

Extraction of natural rubber Natural rubber is stored in the form of a white fluid, called latex inside the bark of rubber trees. Sloppy incisions are made in the bark of the rubber trees. The rubber latex that oozes out from these incisions is collected in small cups. The collected latex is diluted till the percentage of rubber hydrocarbon is reduced to about 15–20. The diluted latex is then passed through sieves (perforated plates) to remove the suspended impurities such as leaves, bark pieces, coagulated rubber, etc.

The crude rubber is separated from the diluted latex by the process called coagulation. Formic acid or acetic acid is being used as the coagulating agent. The purified latex is collected in a large coagulation tank. It is treated with required amounts of formic acid or acetic acid (about 2–4%) till the pH is reduced to about 4.8–5.0. When the rubber is coagulated into a soft white mass, it is called coagulum. The separated coagulum is washed and treated further to produce crepe rubber, smoked rubber and Gutta percha.

Crepe rubber It is a crude form of natural rubber produced by adding sodium bisulphite ($NaHSO_3$) to the coagulum. Sodium bisulphite prevents the oxidation, discoloration and softening of the rubber. The coagulum is allowed to drain for about 2 hours. It is then passed through a "creeping machine" which consists of two rollers with longitudinal grooves. Water is continuously sprayed over the rollers. This helps to wash the rubber sheets and prevents them from sticking to the surface of rollers. The rubber sheets so obtained have an uneven rough surface resembling crepe paper and hence called "crepe rubber". The sheets are finally dried in air for about two weeks.

Smoked rubber It is a type of crude natural rubber, which can be prepared by carrying out the coagulation of the latex in long tanks (1 m wide and 30 cm deep) fitted with partition plates. The diluted latex is poured into the tank, with the partition plates removed. To this, formic acid or acetic acid (2–4%) is added and stirred. The partition plates are then inserted into the tank and the mixture is allowed to stand undisturbed for about 16 hours. The thick slabs of crude rubber formed between the partition plates are removed and washed. The slabs are then passed through a series of smooth rollers of decreasing clearance. Water is continuously sprayed over the rollers. The final roller has a clearance of such a design so as to give ribbed-pattern rubber sheets. The ribbed-pattern on the surface of rubber sheets facilitates quick drying and also prevents the sheets from adhering together on stacking. The sheets are finally dried in a smoke house at 50°C for about 7–10 days. The antiseptic action of smoke prevents the bacterial decompostion of rubber and improves the storage life of rubber. The crude rubber so obtained is amber in colour and is called "smoked rubber".

Gutta percha   Gutta percha is obtained from the matured leaves of *Dichopsis gutta* and *Palaguim gutta* trees. It is recovered by the solvent-extraction, which makes it free from in soluble resins and gums. Chemically, it is known as *trans*-polyisoprene. Gutta percha is hard and tough at room temperture, but becomes soft and tacky when heated to about 100°C. It is soluble in aliphatic hydrocarbon solvents, but insoluble in aromatic and chlorinated hydrocarbon solvents. It absorbs water to a lesser extent than that of natural rubber (Hevea rubber).

Gutta percha is used in the manufacture of toys, golf ball covers, cables, adhesives, etc.

Deficiencies of natural rubber   The pure natural rubber is as useless as pure gold. It is usually not suitable for any useful applications because of its many inherent deficiencies as listed below.

1. It is hard and brittle at low temperatures and soft at high temperatures.
2. It has a low tensile strength.
3. It is readily attacked by acids, alkalis and oxidizing agents.
4. It swells or dissolves in many organic solvents.
5. It has a large water-absorption tendency.
6. It is sensitive to oxidative degradation.
7. It deteriorates in UV and sunlight, and becomes soft and sticky.
8. It is elastic over a limited range of temperatures (10–60°C) and does not regain its original shape after being stretched.
9. It possesses a high tackiness and hence is difficult to handle.

Vulcanization of rubber   Natural rubber is soft and sticky at high temperatures but becomes hard and brittle at low temperatures. It has no useful mechanical and elastic properties. Pure rubber has no practical applications because of its inherent deficiencies. However, it can be transformed into a versatile material having desired properties such as elasticity, toughness and mechanical properties by a process called vulcanization. In 1939, Charles Goodyear accidentally discovered that addition of sulphur to hot rubber causes a drastic improvement in its properties.

Vulcanization is the process of transformation of raw rubber from a plastic state to an elastic state.

Vulcanization is carried out by heating raw rubber with vulcanizing agents like sulphur, sulphur monochloride, hydrogen sulphide, zinc oxide and magnesium oxide. Sulphur is the oldest and most widely used vulcanizing agent. Addition of substances like mercapto benzothiazole, phenyl methyldithiocarbamic acid, benzothiazyl disulphide, increases the rate of vulcanization and hence these substances are used as accelerators.

The raw rubber is mixed with sulphur (3–10%), accelerators and other additives in a rubber mill. The mixture is then heated at 120–150°C. Sulphur acts as a cross-linking agent between the linear polymeric chains of rubber not only at the site of double bonds but also at activated allylic positions.

Vulcanization thus serves to stiffen the material by a sort of anchoring and consequently preventing intermolecular movement of rubber chains. The degree of stiffness introduced in vulcanized rubber depends on the amount of sulphur added. For instance, a tyre rubber may contain 3–5% sulphur, but ebonite rubber (hard rubber) may contain as much as 30% sulphur. To prolong the life of vulcanized rubber antioxidants such as aromatic amines, quinones, phenols, etc. are added during vulcanization.

Synthetic Rubbers

The term synthetic rubber is a misnomer. The synthetic rubbers obtained differ in chemical composition and properties from those of natural rubber. A synthetic rubber is any vulcanizable man-made rubber-like polymers, which can be stretched at least twice its length, but regains its original shape and dimension as soon as the stretching force is released.

Many of the synthetic rubbers are superior to natural rubbers with reference to their improved properties such as better resistance to deterioration by oil, organic solvents, oxidizing agents, heat and sunlight, etc.

Neoprene rubber (polychloroprene)   Neoprene is produced by the emulsion polymerization of chloroprene (2-chloro-1, 3-butadiene) in the presence of aluminium persulphate at about 40°C. Chloroprene is usually emulsified using soap such as rosin soap.

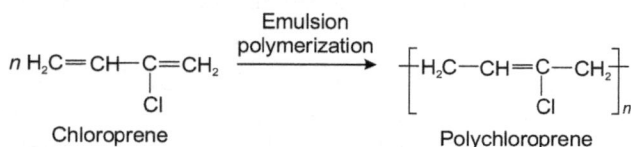

$$n \ H_2C{=}CH{-}\underset{\underset{Cl}{|}}{C}{=}CH_2 \quad \xrightarrow[\text{polymerization}]{\text{Emulsion}} \quad \left[\!H_2C{-}CH{=}\underset{\underset{Cl}{|}}{C}{-}CH_2\!\right]_n$$

Chloroprene   Polychloroprene

Neoprene rubber is not generally vulcanized using sulphur, as the reaction is very slow even in the presence of catalysts. However, it is vulcanized using ZnO or MgO. It is resistant to oil and common chemicals. It exhibits high tensile strength and better resistance to oxidative degradation and ageing.

Neoprene is used in making press rollers. For water- or alcohol-based inks, rollers made with neoprene are used. It is used as adhesives and surface coatings.

Buna-S (SBR)   It is a copolymer of 1, 3-butadiene and styrene. Buna-S is manufactured by emulsion polymerization method. Butadiene and styrene are mixed in the ratio 3 : 1 by weight and made into an emulsion in water using soap solution. The polymerization is carried out at 5°C in the presence of benzoyl peroxide as initiator.

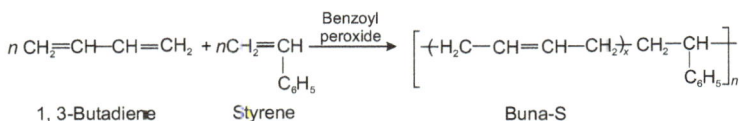

$$n\ CH_2{=}CH{-}CH{=}CH_2 + nCH_2{=}\underset{\underset{C_6H_5}{|}}{CH} \xrightarrow{\text{Benzoyl peroxide}} \left[{+}H_2C{-}CH{=}CH{-}CH_2{\}}_x\ CH_2{-}\underset{\underset{C_6H_5}{|}}{CH}{+} \right]_n$$

1, 3-Butadiene   Styrene   Buna-S

It can be vulcanized with sulphur. It has high abrasion resistance, load-bearing capacity and resilience. It is much superior to natural rubber with reference to ageing and ozone resistance.

Buna-S is used as an adhesive and in the manufacture of coated fabrics. It is used as the binder in coating materials, which are employed in the manufacture of coated papers. It is used in the manufacture of light-duty tyres.

Buna-N (nitrile rubber) Buna-N is a copolymer of 1, 3-butadiene and acrylonitrile (vinyl nitrile). It is produced by emulsion polymerization technique. The monomers, 1, 3-butadiene and acrylonitrile, are held in water as emulsion using soap solution as emulsifier. The polymerization reaction is performed using redox initiator systems such as H_2O_2 and $FeSO_4$ either in cold (5°C) or at warm (30°C) conditions.

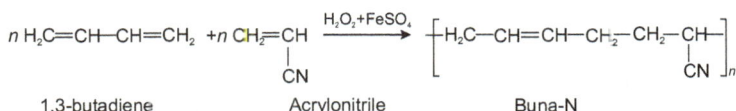

$$n\ H_2C{=}CH{-}CH{=}CH_2 + n\ CH_2{=}\underset{\underset{CN}{|}}{CH} \xrightarrow{H_2O_2+FeSO_4} \left[H_2C{-}CH{=}CH{-}CH_2{-}CH_2{-}\underset{\underset{CN}{|}}{CH}{-} \right]_n$$

1,3-butadiene   Acrylonitrile   Buna-N

It shows excellent resistance to petroleum oils and this property increases with increase in the percentage of acrylonitrile. It exhibits high tensile strength, good heat and abrasion resistance. But it is less resilient than natural rubber.

Buna-N is widely used in the production of press rollers, rubber flexoplates, duplicate plates and lithographic rubber blankets. It is used in making adhesives and oil-resistant foams. It is used for impregnating papers, textiles and leather.

It is used as a binder in coating materials for making coated papers.

Butyl rubber   Butyl rubber is obtained by copolymerization of isobutylene with small amounts of isoprene (1–5%).

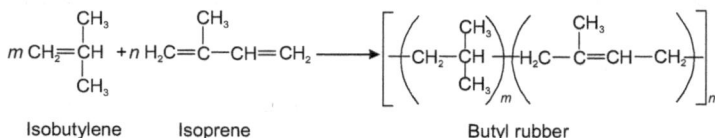

Isobutylene   Isoprene   Butyl rubber

It has low gas permeability, better ozone resistance, and good solvent resistance. It shows excellent resistance to heat, abrasion, ageing and chemicals.

It is used in the production of printing rollers and duplicate plates.

Polysulphide rubbers (thiokols)   They are produced by condensation polymerization of an organic dihalide like ethylene dichloride with an inorganic polysulphide like Na_2S_4.

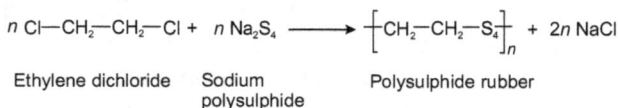

Ethylene dichloride   Sodium   Polysulphide rubber
                       polysulphide

They possess low tensile strength and poor resistance to abrasion. They are resistant to oils, solvents, fuels, ozone and sunlight.

Polysulphide rubbers are used in the production of press rollers for flexography. They are also useful in tyre tread coatings and oil tank linings.

Silicone rubbers   Silicone rubbers (e.g. polydimethyl siloxane) are the most important inorganic polymers. They contain chains of alternating silicon and oxygen atoms and each silicon atoms attached with two alkyl groups.

They are obtained by the condensation polymerization of alkyl silanols (dimethyl silanol), which in turn are obtained by hydrolysis of alkyl chlorosilanes (dimethyl chlorosilane).

$$n \; \text{Cl—Si—Cl} \xrightarrow[-2n\text{HCl}]{\text{Hydrolysis}} n \; \text{HO—Si—OH} \xrightarrow[-n\text{H}_2\text{O}]{\text{Polymerization}} \left[\begin{array}{c}\text{CH}_3 \\ \text{Si—O} \\ \text{CH}_3\end{array}\right]_n$$

| Dimethyl chlorosilane | Dimethyl silanol | Silicone rubber |

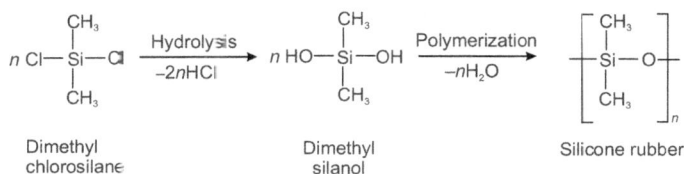

By controlling the degree of polymerization, silicone rubbers may be obtained in the form of fluid, resin or rubber-like material. Because of the absence of unsaturation, they cannot be vulcanized using sulphur. Alternatively they can be vulcanized by using peroxides such as H_2O_2 or benzoyl peroxide. On vulcanization, silicon chains are cross-linked through methylene bridges.

Because of the presence of silicon–oxygen links, they exhibit outstanding stability at high temperatures (up to 250°C) and good water resistance. They show exceptional resistance to prolonged exposure to sunlight, weathering, oils, acids and alkalis. They remain flexible over a wide temperature range (90–250°C).

Silicone rubbers are extensively used in waterless lithographic printing. They provide an ink-repellent surface in waterless lithoplates. In the waterless lithographic plate, the ink-repellent silicone rubber surface is produced by the condensation cross-linking of a linear silanol in the presence of UV light. They have the property of improving the shipping power of anything to which they are applied, such as collating machines, the feed boards and back cylinders of presses. They minimize ink transfer from sheets to press parts such as delivery tapes and feed-board wheels. The dimethyl silicones are used in food-packaging inks.

Laminated Plastics

Laminated plastics are the reinforced plastic articles. They are prepared by impregnating a resin on the sheets of paper, cloth, wood, glass, asbestos fibres, nylon or metal. Thermosetting resins like phenol–formaldehyde, urea–formaldehyde, epoxy resins and silicones are most commonly used for this purpose.

The reinforcing sheets (paper, cloth, wood, etc.) are passed through a solution of thermosetting resin (dissolved in a suitable solvent like alcohol). The impregnated sheets are then dried well below the curing temperature of the resin. The dried sheets are trimmed and piled over one another to the required thickness. The piled sheets are then subjected to curing in a hydraulic press by the application of heat (130–180°C) and pressure (120 kg/cm^2). The curing time varies with the type of materials used and the thickness of the laminates required. On curing, the resin becomes hard and bonds the sheets to form a compact material.

The important properties of laminated plastics are light weight, high tensile strength, excellent electrical insulation, good impact and chemical resistance. They are impermeable to water and oils.

They are used as base materials for making printing blocks. Laminates are widely used in packaging. They are extensively used for decorative purposes such as panelling for walls and automobiles, and for tabletops. Laminates are also used in satellites, boat hulls, etc.

REVIEW QUESTIONS

1.   What are polymers? How are they classified?
2.   Explain the following terms.
  i.   Monomer
  ii. Polymerization
  iii. Functionality
  iv. Degree of polymerization
  v.   Copolymer

3. Distinguish between
    i.   Addition polymerization and condensation polymerization
    ii. Thermoplastics and thermosetting resins
    iii. Addition polymerization and copolymerization

4. What are addition and condensation polymers? Give examples.

5. Write down the names and structural formulae of monomer/s of the following polymers.
    i.   Polyvinyl chloride
    i.   Glyptal resin
    iii. Epoxy resin
    iv.   Buna-S
    v.   Nylon 6,6
    vi.   Polyurethane
    vii. Thiokol
    viii. Neoprene
    x.   Silicone rubber

6. How are the following polymers produced? Give their applications in printing industry.
    i.   Polypropylene
    ii. Polyvinyl chloride
    iii. Polyvinyl acetate
    iv. Nylon 6,6
    v.   Alkyd resin
    vi. Epoxy resin
    vii. Polyurethane

7. Give the preparation, properties and uses of phenol–formaldehyde resin.

8. Describe the synthesis and applications of terylene.

9. Write notes on:
    i. Plasticizers
    ii. Copolymerization

10. What are elastomers? Write the structure of natural rubber and gutta percha.

11. How is natural rubber obtained from latex?

12. Explain the preparation of the following.

    i. Crepe rubber

    ii. Smoked rubber

    iii. Gutta percha

13. What are the drawbacks of natural rubber?

14. Write a note on vulcanization of rubber.

15. Write down the names and structures of polymers prepared from the following monomers.

    i. Styrene

    ii. Vinyl acetate

    iii. Ethylene glycol and terephthalic acid

    iv. Hexamethylene diamine and adipic acid

    v. Bisphenol A and epichlorohydrin

    vi. Bisphenol A and diphenyl carbonate

16. Give the preparation and applications (in printing industry) of the following.

    i. Buna-N

    ii. Butyl rubber

    iii. Buna-S

    iv. Thiokol

    v. Neoprene

17. Explain the preparation, properties and applications of silicone rubber.

18. Discuss the preparation, properties and uses of laminated plastics.

19. Justify the following statements.

    i. Rubber needs vulcanization.

    ii. Thermosetting plastics cannot be remoulded.

    iii. All simple molecules are not monomers.

    iv. PVC is usually plasticized before use.

ADDITIONAL READING

Billmeyer, F. W., Jr. (1984). *Text Book of Polymer Science*, 3rd edn. Wiley-Interscience, New York.

Ferry, M.H. and Becker, A.V. (2004). *Hand Book of Polymer Science and Technology*, 1st edn. Vol. 1 & 2. CBS Publishers, New Delhi.

Gowariker, V.R., Vishwanathan, N.V. and Sreedhar, J. (1996). *Polymer Science*. New Age International (P) Ltd., New Delhi.

Harper, C.A. (ed.). (1996). *Hand Book of Plastics, Elastomers and Composites*, 3rd edn. McGraw-Hill Professional Book Group, New York.

Mark, H.F., Bikales, N. and Kroschwitz, J.J. (2003). *Encyclopedia of Polymer Science and Technology*, 3rd edn. Vol.1–4. John Wiley and Sons, New York.

Odian, G. (2004). *Principle of Polymerization*, 4th edn. John Wiley and Sons, Hobokon, NJ.

5 COLLOIDS

INTRODUCTION

Thomas Graham (1861) investigated the diffusion of substances in solution through parchment membrane and based on the results obtained, he classified the substances into two classes.

1. *Crystalloids* The substances like sugar, salts, acids and bases which generally exist in crystalline state and diffuse readily through the parchment membrane are termed as crystalloids.

2. *Colloids* The amorphous substances like gelatin, gum, glue, and albumin which diffuse at a very slow rate are given the name colloids.

These fundamental observations made the landmark in the development of a new branch of science known as **colloid science**.

Although the work of Graham was of fundamental importance, it was soon realized by many investigators that the distinction between crystalloids and colloids was not rigid. Because every substance, irrespective of its nature, can be converted to the colloidal state under suitable conditions. For example, sodium chloride is a crystalloid in water but behaves like a colloid in benzene. Soap is colloidal in water and crystalloid in alcohol.

Ostwald and Von Weimarn have proposed the first rational classification of colloids. They introduced the term "dispersed system" and the size of particles was taken as the main criterion in classifying the colloids. The dispersed systems may be divided into three classes.

i. **True systems**   The system in which the particles have the size less than 1 mμ (10 Å). It is a homogeneous system and the dispersed phase contains atoms, molecules, or ions.

ii. **Colloid dispersed systems**   The system in which the particles have the size range 1 mμ–200 mμ (10Å–2000 Å) and are invisible under ordinary microscope. It is a heterogeneous (two-phase) system.

iii. **Coarse dispersed systems (suspension)**   The system in which the size of the particles is more than 200 mμ (2000 Å) and thus settle under the influence of gravity. The particles are neither dissolved nor dispersed but are suspended in the medium in a heterogeneous manner. The particles are visible to the naked eye.

COLLOIDAL SYSTEMS

A colloidal system is a heterogeneous two-phase system in which a substance is dispersed in colloidal state in an insoluble medium.

The medium in which the dispersion takes place is called the dispersion medium or continuous phase. The substance dispersed in the medium is called the dispersed phase or discontinuous phase. The continuous phase usually forms the larger fraction of the colloidal system. Both the dispersed phase and dispersion medium may be a solid, liquid or gas.

The colloidal dispersions are thermodynamically unstable and tend to coagulate or precipitate on standing unless suitably stabilized. In order to stabilize the colloidal dispersion, it is essential to add a stabilizing agent which keeps the colloidal particle apart thus preventing the coagulation.

TYPES OF COLLOIDAL SYSTEMS

Based on the state of aggregation of the dispersed phase and dispersion medium, the colloidal system may be classified into nine types (Table 5.1).

Table 5.1 Different types of colloidal systems

Dispersed phase	Dispersion medium	Name	Examples
Solid	Solid	Solid sol	Gems, ruby glass, minerals, coloured precious stones, pigmented plastics, etc.
Solid	Liquid	Sol	Glue, Indian ink, starch, gold, AgCl, As_2S_3 dispersed in water, muddy water, etc.
Solid	Gas	Solid aerosol	Smoke, dust, etc.
Liquid	Solid	Gel	Jellies, curd, cheese, butter, etc.
Liquid	Liquid	Emulsion	Milk, cream, oil-in water, water-in-oil, etc.
Liquid	Gas	Liquid aerosol	Fog, mist, cloud, etc.
Gas	Solid	Solid foam	Pumice stone, cork, adsorbed gases, rubber, cake, bread, etc.
Gas	Liquid	Foam	Froth, whipped cream, soap lather, etc.

A colloidal system of two gases is not possible, since they are completely miscible and they do not form the two phases.

In general, when the dispersion medium is a gas the colloidal system is known as **aerosol**. If a solid is present in dispersed condition in a liquid medium, the system is called **sol**. When the dispersion medium is water, the sol is called as hydrosol or aquasol. If the dispersion medium is alcohol or benzene or any other organic liquid, the system is termed as alcosol, benzosol and organosol, respectively.

SOLS

Sols are the colloidal systems in which solid particles are dispersed in a liquid medium. Generally, the term colloid is referred to the sol.

Types of Sols

Based on the affinity of the dispersed phase for the dispersion medium, sols are subdivided into two types.

1. Lyophobic (solvent-hating) sols
2. Lyophilic (solvent-loving) sols

Lyophobic sols are those dispersions in which the dispersed phase has very little affinity for the dispersion medium. Some typical examples are metals, As_2S_3, $Fe(OH)_3$, and AgCl dispersed in water.

Lyophilic sols are those in which the dispersed phase has great affinity for the dispersion medium. Some typical examples are gelatin, gum, glue and starch dispersed in water.

When the dispersion medium used is water, then the two types of sols are particularly termed as hydrophobic and hydrophilic sols. The lyophilic sols readily pass into the colloidal state when brought in contact with the dispersion medium, but not the lyophobic sols. The lyophobic colloids are much less stable and the residue left on evaporation cannot be reconverted into sol by ordinary means. Therefore, they are said to be irreversible colloids. Lyophilic colloids are more stable. They are also known as reversible colloids, since the residue obtained on evaporation of the dispersion medium can be reconverted into the sol on mere addition of the liquid medium.

The main differences between the two types of sols are tabulated in Table 5.2.

Table 5.2 Differences between lyophobic and lyophilic sols

Property	Lyophobic sols	Lyophilic sols
Preparation	They are prepared with difficulty and require special methods.	They are formed very easily by shaking or warming the substance with water.
Reversibility	They are irreversible colloids.	They are reversible colloids.
Stability	The particles are less stable.	The particles are more stable.
Surface tension	Same as that of the medium.	Lower than that of the medium.
Viscosity	Same as that of the medium.	Much higher than that of the medium.
Visibility	The particles can be easily detected under an ultramicroscope.	Particles cannot be readily detected under the ultramicroscope.
Effect of an electric field	The particles can migrate towards cathode or anode.	The particles may or may not migrate.
Action of electrolytes	Addition of a small quantity is sufficient to cause coagulation.	Large amount is needed to cause coagulation.
Solvation	The particles are less solvated.	The particles are highly solvated.

Preparation of Sols

The lyophilic sols can be readily prepared by shaking or warming the substances like starch, gelatin, gum arabic, etc. with suitable liquids (usually water). The sols obtained are of reversible type and are thermodynamically stable. But lyophobic sols require some special methods for their preparation. The primary consideration in the preparation of sols is that the dispersed particles should be within the colloidal range (1 mμ–200 mμ). Thus the preparation of sols involves the production of particles of colloidal dimensions. The different methods used for the preparation of lyophilic sols are broadly grouped into two types.

Dispersion methods   In this method, larger particles of the substances are broken into smaller particles of colloidal dimension in the presence of dispersion medium.

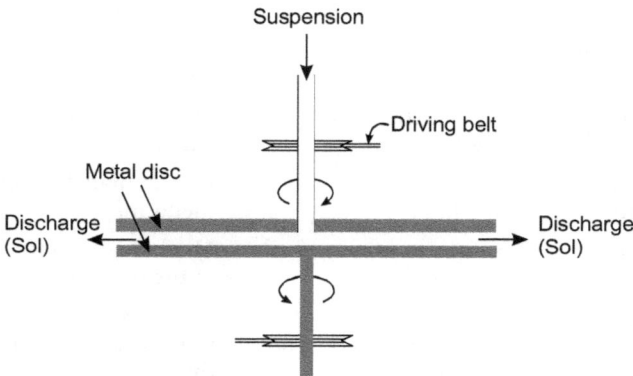

Figure 5.1   Colloid mill

Mechanical grinding   In this method, the substance to be dispersed is finely subdivided by grinding. The grinding is done in a special type of mill known as **colloid mill** (Figure 5.1). It consists of two metal discs held at a small distance apart from each other and capable of rotating at high speed (about 7000 rpm) in opposite directions. Actually the distance between the two discs controls the size of the colloidal particles obtained. The substance is finely ground by usual methods and agitated with the dispersion medium to produce a suspension. It is then fed into the colloid mill in between the rotating discs. The suspension particles are torn off to the colloidal dimensions. A stabilizing agent is then added to get a stable sol.

Black ink (Indian ink) is prepared by grinding the lamp black with water in the presence of a little gum arabic as stabilizing agent. Paints, varnishes, dental creams, dyestuff, and ointments are also prepared by this method.

Bredig's arc method (electrodispersion method) This method is particularly used to prepare sols of metals such as silver, gold, platinum, copper, etc.

An electric arc is struck between two electrodes of a metal (like gold, silver, or copper) in water containing traces of alkali as stabilizer (Figure 5.2). It is believed that by the heat of the spark, the metal first changes into vapour which then condenses in water to give aggregates of colloidal range.

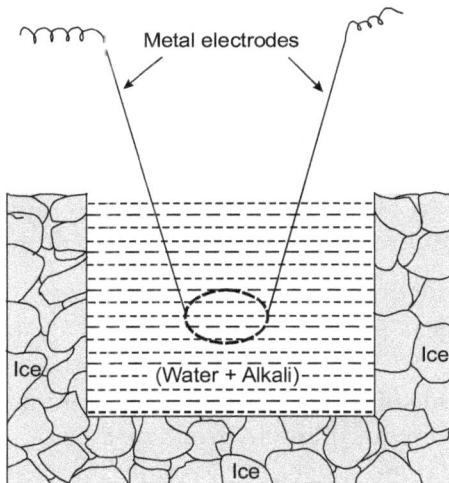

Figure 5.2 Bredig's arc method

Svedberg (1906) modified this method and prepared sols in non-aqueous medium (like diethyl ether) by striking an arc with high-frequency alternating current which greatly reduces the decomposition of the medium.

Peptization The process of conversion of a coarse precipitate into colloidal state by the addition of small amount of an electrolyte is called peptization. The electrolyte added is known as peptizing agent.

For example, a freshly formed precipitate of ferric hydroxide can be converted into colloidal state by adding a small amount of dilute solution of ferric chloride. Stannic oxide precipitate can be converted into a stable sol by adding a small amount of dilute hydrochloric acid. A freshly formed AgCl suspension is peptized by shaking it with a very dilute solution of silver nitrate.

Condensation methods   In this method, the molecules or ions in a true solution are made to aggregate to form particles of colloidal dimensions by suitable chemical or physical process.

Double decomposition   Arsenic sulphide sol is prepared by passing hydrogen sulphide gas through a cold dilute solution of arsenious oxide in water.

$$As_2O_3 + 3H_2S \longrightarrow \underset{(Sol)}{As_2S_3} + 3H_2O$$

Similarly, a sol of Prussian blue may be prepared by mixing very dilute solutions of ferric chloride and potassium ferrocyanide.

$$3K_4Fe(CN)_6 + 4FeCl_3 \longrightarrow \underset{(Prussian\ blue\ sol)}{Fe_4[Fe(CN)_6]_3} + 12KCl$$

Silicic acid sol is obtained by mixing dilute solutions of sodium silicate with dilute hydrochloric acid.

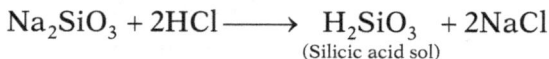

$$Na_2SiO_3 + 2HCl \longrightarrow \underset{(Silicic\ acid\ sol)}{H_2SiO_3} + 2NaCl$$

Reduction   The sols of metals (like gold, silver, and platinum) in water are obtained by reducing the dilute aqueous solutions of their salts with suitable reducing agents (such as formaldehyde, phenyl hydrazine, tannic acid, etc.)

Gold sol may be obtained by boiling 3–4 ml of 0.2N K_2CO_3 solution and 1 ml of 1% $AuCl_3$ solution in about 150–200 ml of distilled water, followed by gradual addition of 1–3 ml of 0.3% formaldehyde solution with continued stirring.

$$2AuCl_3 + 3HCHO + 3H_2O \longrightarrow \underset{(Sol)}{2Au} + 3HCOOH + 6HCl$$

The silver sol can be prepared by reducing silver nitrate solution by tannic acid.

Hydrolysis   The sols of many oxides and hydroxides of iron, aluminium, tin, etc. are prepared by hydrolysis of their salts in aqueous solution.

For example, ferric hydroxide sol is obtained by adding freshly prepared saturated solution of ferric chloride dropwise to an excess of boiling water.

$$FeCl_3 + 3H_2O \longrightarrow \underset{(Sol)}{Fe(OH)_3} + 3HCl$$

Oxidation   The sols of some non-metals like sulphur or iodine are prepared by oxidation. For example, sulphur sol is prepared by passing hydrogen sulphide through a solution of sulphur dioxide in water.

$$2H_2S + O_2 \longrightarrow \underset{(Sol)}{2S} + 2H_2O$$

Sulphur sol can also be obtained by bubbling oxygen through a solution of hydrogen sulphide in water.

$$2H_2S + O_2 \longrightarrow \underset{(Sol)}{2S} + 2H_2O$$

Sol of iodine is obtained by oxidizing hydroiodic acid (aqueous solution of HI) with iodic acid (HIO_3).

$$5HI + HIO_3 \longrightarrow \underset{(Sol)}{3I_2} + 3H_2O$$

By change of solvent   The sols of a number of substances can be prepared by taking a solution of the substance in one solvent and pouring it into another solvent in which the substance is insoluble. Thus, when a saturated solution of sulphur in alcohol is poured into water, colloidal sol of sulphur is obtained. Phosphorus sol may be prepared similarly.

If a solution of AgI in an aqueous solution of KI is poured into an excess of water, the complex ion $[AgI_2]^-$ decomposes into AgI and I^- ion. The AgI obtained is in the form of a colloidal sol.

Purification of Sols

The presence of large amounts of electrolyte impurities causes the precipitation or coagulation of the sols. However, the presence of a small amount of an electrolyte is necessary for the stability of a sol. Therefore, it is necessary to eliminate the excess amount of electrolytes. The purification of sols is generally performed using the following methods.

Dialysis The porous membranes (such as parchment membrane or cellophane paper) are permeable to molecules of solvent and low-molecular-weight solutes but not to the colloidal particles. This is because of the larger size of the colloidal particles. Whenever such a membrane separates colloidal sols from distilled water, it permits the passage of only the electrolytes through it but not the colloidal particles. The method of dialysis is based on this principle. The process of separating a crystalloid from a colloid by diffusion through a porous membrane is known as dialysis. The apparatus used for this purpose is known as dialyser.

Graham's dialyser consists of a cylinder open at both ends, and one end of which has a membrane attached over it. The sol to be dialysed is filled in the cylinder which is then suspended in a large dish containing distilled water (Figure 5.3).

The distilled water is renewed from time to time. But a continuous flow of water is preferred, as such a flow generally accelerates dialysis. Ordinary dialysis is a slow process and takes several hours for completion.

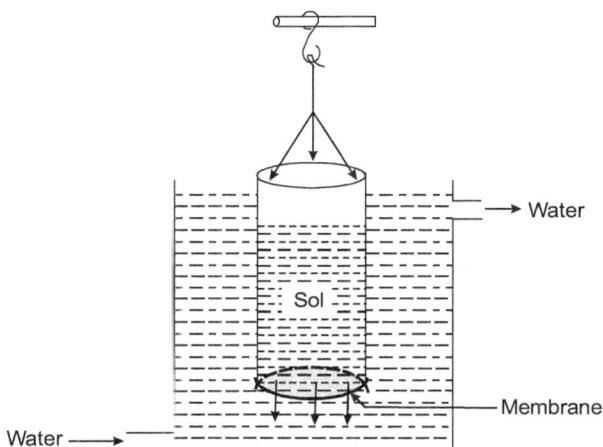

Figure 5.3   Dialysis of sols

Electrodialysis   The rate of dialysis can be greatly improved under the influence of electric field.

The sol is placed in the middle compartment between the two dialysing membranes while pure water is circulated into the outer compartments of the dialyser (Figure 5.4).

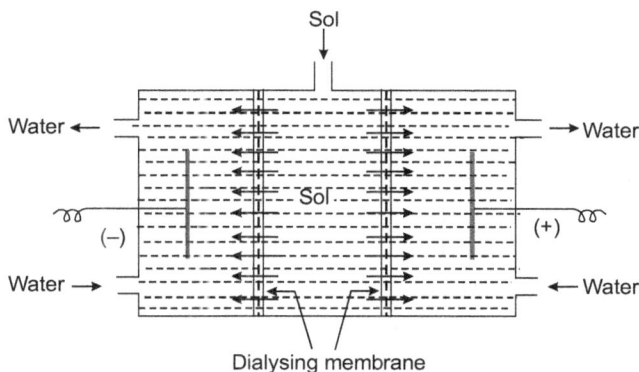

Figure 5.4   Electrodialysis

There are two electrodes fixed, one in each outer compartment. On applying a suitable emf, the ions of the electrolyte diffuse through the membrane and migrate towards the oppositely charged electrodes, leaving behind the colloidal particles in the middle compartment.

Ultrafiltration   The pores of ordinary filter paper are large enough for the colloidal particles (size less than 200 mμ) to pass through, so they cannot be employed for filtering sols. But if the pores are made smaller, the colloidal particles may be retained on the filter paper. For this, the ordinary filter paper is soaked in a solution of gelatin or colloidin (in acetic acid) and then hardened by soaking in formaldehyde. The filter papers so obtained are known as ultrafilters. The process of separating colloids from solutes by using ultrafilters is known as ultrafiltration. The pores in the ultrafilters will be large enough to permit the passage of molecules and ions in true solutions, but will be small enough to retain the colloidal particles. Graded filters with different pore sizes can be made by varying the concentration of gelatin. By using a series of graded ultrafilters it may even be possible to separate colloidal particles of different sizes. Since the ultrafiltration proceeds very slowly, pressure or suction is applied to speed up the process. By this process it is possible to separate the colloidal particles in the form of slime, from the media containing the electrolytes. These slimes may then be suspended in pure media to get the sol.

Properties of Sols

General properties

Heterogeneous nature   True solutions are homogeneous in character but colloidal particles, being larger, can form only heterogeneous system consisting of two phases, dispersed phase and dispersion medium.

Filterability   The colloidal particles can readily pass through ordinary filter papers but not through ultrafilter papers. This is because the size of colloidal particles is much less than the size of pores in an ordinary filter paper and greater than that in the ultrafilters.

Colligative properties The magnitude of colligative properties (i.e., osmotic pressure, lowering of vapour pressure, elevation of boiling point and depression in freezing point) depends upon the number of solute particles present in a given amount of the solvent. Now, the colloidal particles are aggregates of simple molecules. For example, in the case of arsenic sulphide sol, each particle is composed of about 1000 molecules. Hence, for a given amount of arsenic sulphide sol, the number of particles in the sol will be only one-1000th of the number present in a true solution of similar concentration. Therefore, the colloidal systems give low osmotic pressure, elevation of boiling point and depression in freezing point at about the same temperature as pure dispersion medium.

Kinetic properties

Diffusion The colloidal particles in a colloidal dispersion diffuse from a region of high concentration to a region of low concentration until the concentration is uniform throughout. However, the rate of diffusion is very slow compared to the particles in true solutions.

Brownian movement When a colloidal dispersion is observed through an ultramicroscope, the colloidal particles are found to be in a state of continuous, rapid and random motion. This phenomenon was first observed by Robert Brown (an English botanist in 1827) and hence, is called the Brownian movement.

Brownian movement is due to the bombardment of colloidal particles by molecules of dispersion medium which are in constant motion. The colloidal particles are relatively heavier than the molecules of the dispersion medium. Due to the bombardment, the particles gain kinetic energy equal to that of the medium and start moving randomly. Brownian movement is more rapid when the colloidal particle is smaller and the medium is less viscous. It increases with an increase in temperature.

The Brownian movement is not observed in ordinary suspensions, because the mass of each particle in this case is so large that the bombardment of the molecules of the medium produces a little effect on them.

Sedimentation   Although colloidal dispersions are stable over a long period of time, they, however, tend to settle slowly under the influence of gravity on prolonged standing. This phenomenon is called sedimentation.

Optical properties

Tyndall effect   When a beam of light is passed through a true solution, the path of light cannot be seen unless the eye is placed directly in its path. However, when a beam of light is passed through a colloidal dispersion it becomes visible as a bright streak. This phenomenon is known as **Tyndall effect** and the illuminated path is known as **Tyndall cone**. Tyndall effect is due to the scattering of light from the surface of the colloidal particles. The scattering of light cannot be due to the simple reflection, because the size of the particles is smaller than the wavelength of the visible light which are unable to reflect light waves. Thus, Tyndall effect is observed only when the following conditions are satisfied.

i.   The diameter of the dispersed particles is not much smaller than the wavelength of light used.

ii.   The refractive indices of the dispersed phase and the dispersion medium must differ greatly in magnitude. This condition is fulfilled by lyophobic colloids. Therefore, Tyndall effect is more pronounced in lyophobic sols and is weak in lyophilic sols.

True solutions do not show the Tyndall effect. Therefore, Tyndall effect is useful to distinguish between colloidal and non-colloidal systems.

Ultramicroscope is used to observe the Tyndall effect. A powerful beam of light from an arc lamp is condensed by a system of lenses and is passed through the colloidal solution.

The scattered beam is viewed through a microscope placed at right angles to the beam (Figure 5.5).

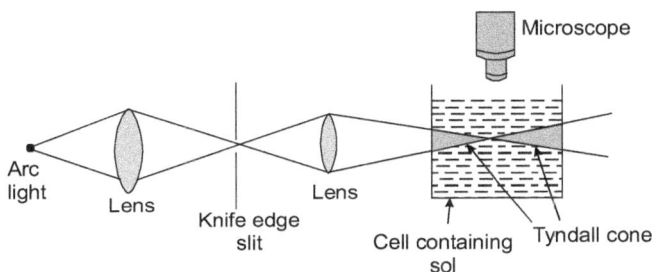

Figure 5.5 Ultramicroscope

The background of the field is dark and the colloidal particles are detected as bright spots of light against a dark background. It may be noted that we do not see the actual particles which are too small to be visible. We only see the light scattered by them.

Electrical properties The colloidal particles are electrically charged. All the particles in a given colloidal system carry the same charge and the dispersion medium has an equal but opposite charge. As a result, the colloidal system as a whole is electrically neutral. The presence of similar charges on colloidal particles is mainly responsible for the stability of the system. The colloidal particles with similar charges on them repel each other and do not combine to form larger particles by aggregation, when they come close to one another. Metal hydroxide sols (like $Fe(OH)_3$, $Al(OH)_3$, etc.) and some metal sols (such as Bi, Pb, Fe, etc.) have positively charged particles. Metal sulphide sols (such as As_2S_3, CdS, Sb_2S_3, etc.) and many metal sols (such as Ag, Au, Pt, etc.), Prussian blue, silicic acid, tannic acid, gelatin, starch sols, etc. are negatively charged.

The electrical properties of colloidal particles can be explained satisfactorily by assuming the existence of an electrical double layer of opposite charges (Helmholtz double layer) at the solid–liquid interface. A layer of charge on the colloidal particle is surrounded by another layer of ions of opposite kind on the dispersion medium side (Figure 5.6). The layer of charge on the colloid particles is the fixed part of the double layer and consists of either positive or negative ions.

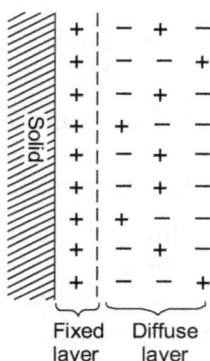

```
        +  |  −  +  −
        +  |  −  −  +
        +  |  −  +  −
   S    +  |  +  −  −
   o    +  |  −  +  −
   l    +  |  −  +  −
   i    +  |  +  −  −
   d    +  |  −  +  −
        +  |  −  −  +
        └─┘ └────────┘
       Fixed   Diffuse
       layer   layer
```

Figure 5.6   Electrical double layer

The other layer extends to the medium and is known as the mobile or diffuse layer. This layer consists of both types of ions but its net charge is equal and opposite to that on the fixed part of the double layer. The potential difference across the double layer existing between the surface of the colloidal particles and the medium of dispersion is called zeta potential or electrokinetic potential. When an electric field is applied to a colloidal solution, the colloidal particles and the medium will migrate in opposite directions. If the experiment is arranged such that the colloidal particles can move but not the medium, the phenomenon is known as electrophoresis. On the other hand, when only the medium is allowed to migrate and not the particles, the phenomenon is known as electro-osmosis.

Electrophoresis   The migration of the electrically charged colloidal particles and the medium towards the oppositely charged electrodes under an applied electric field is called electrophoresis. This effect can be studied in the apparatus called Burton tube (Figure 5.7). The apparatus consists of a U-tube provided with a stopcock through which it is connected to a funnel-shaped reservoir. A small quantity of water is taken in the U-tube and a reasonable quantity of the sol is taken in the reservoir. The stopcock is then slightly raised so as to introduce the sol into the U-tube by upward displacement of water, producing boundary in each arm. A voltage of 50–200 volts is then applied using two platinum electrodes which are immersed only in the water layer. The movement of the particles can readily be followed by naked eye or using a lens.

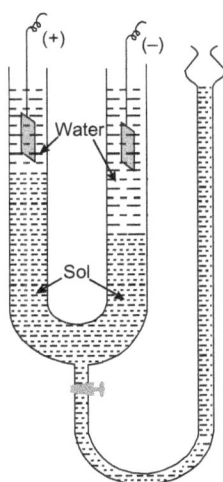

Figure 5.7   Burton tube

When the colloidal particles are negatively charged (such as in As_2S_3 sol), the level of the sol falls gradually on the negative electrode side and rises simultaneously on the positive electrode side. When the colloidal particles are positively charged, the reverse effect is observed. Thus, by noting the direction of motion of the particles in an electric field, it is possible to determine the sign of the charge carried by the particles.

Electrophoresis can also be utilized for quantitative measurement of the rate of migration of colloidal particles in the applied field. The electrophoretic mobility of colloidal particles is defined as the distance travelled (in cm) by them in one second under a potential gradient of one volt per centimetre.

Different colloidal species migrate at different rates; therefore a mixture of colloidal systems can be separated by electrophoresis. Polysaccharides, proteins and nucleic acids can be fractionated by electrophoresis.

Electro-osmosis   When electrophoresis of colloidal particles is prevented by some suitable means, the dispersion medium can be made to move under the influence of an electric field. This phenomenon is called electro-osmosis.

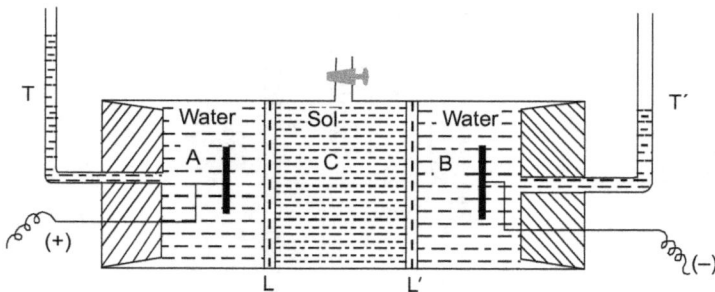

Figure 5.8   Electro-osmosis

The simple apparatus used for electro-osmosis is shown in Figure 5.8. The sol is placed in the middle compartment C, which is separated from the compartments A and B filled with water by the dialysing membranes L and L´. The water-containing compartments A and B are also provided with two side tubes T and T´ to observe the level of water.

The membranes prevent the movement of the colloidal particles. When a potential difference is applied by inserting two electrodes in the compartments A and B, the water will begin to move. The direction of flow of water depends on the nature of charge on the colloidal particles. For positively charged sols, the medium (water) is negatively charged and hence, the flow of water will take place from the compartments B to A. Hence the level of water in the side tube T would be seen to rise. If the sol is negatively charged, the flow of water will take place from the compartment A to B, and the level of water on the side tube T´ would start rising.

The Origin of Charge on Colloidal Particles

It has been observed that sols are invariably associated with small quantities of electrolytes and that if the latter are completely removed (by dialysis), the sols become unstable. It is believed, therefore, that the charge on the colloidal particles is due to the preferential adsorption of either positive or negative ions of the electrolytes on their surface. If the particles have a preference to adsorb negative ions, they acquire a negative charge and vice versa. For example, the negative charge on arsenic sulphide sol is due to the preferential adsorption of sulphide ions on the surface of the particles. The sulphide ions are furnished by the ionization of hydrogen sulphide which is present in traces. Similarly, the negative charge on metal sols prepared by the Bredig's arc method is due to the adsorption of OH^- ions furnished by the traces of alkali added. Likewise, the positive charge on ferric hydroxide sol prepared by the hydrolysis of ferric chloride is due to the preferential adsorption of Fe^{3+} ions on the surface of the particles.

An interesting case is observed in the preparation of silver chloride sol. If a dilute solution of $AgNO_3$ is added to a dilute solution of KCl, a negatively charged sol of silver chloride is formed. This is due to the preferential adsorption of Cl^- ions. But, if a dilute solution of KCl is added to a slight excess of a dilute solution of $AgNO_3$, a positively charged sol of silver chloride is formed due to the preferential adsorption of silver ions. When two or more ions are present in the dispersion medium, the ion which is more nearly related chemically to the colloidal particle is preferentially adsorbed by it. Thus in ferric hydroxide sol, it prefers Fe^{3+} ions and not Cl^- ions.

Another possible way in which colloidal particles acquire charge sometimes is by direct ionization of the material constituting the particles. Examples are acidic and basic dyestuff. Acidic dyestuff ionizes yielding H^+ ions in solution, thereby leaving an equivalent amount of negative charge on the particles (Figure 5.9a). Basic dyestuff, on the other hand, ionizes yielding OH^- ions in solution and an equivalent amount of positive charge on the particles (Figure 5.9b).

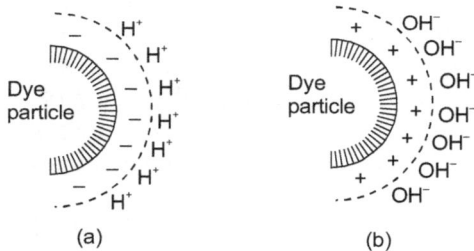

(a)   (b)

Figure 5.9 Origin of charge in dyestuff (a) Acidic (b) Basic

Stability of Sols

In the case of lyophobic colloids, the stability is due to their electric charge, but in the case of lyophilic colloids, it is due to their electric charge and extensive solvation. To bring about precipitation of colloids, the particles must coalesce into aggregates that are large enough to settle out. Since all the colloidal particles are of same charge, they repel each other when they approach close to each other. The magnitude of the repulsive forces depends upon the magnitude of the surface charge and the thickness of the electrical double layer. Hence, as long as the charges are present, each particle is protected from every other by an electric field, and approach close enough for coalescence into larger aggregates is difficult. Thus, the charge on the particles is responsible for the stability of lyophobic sols.

In the case of lyophilic sols, the stability is due to electric charges as well as the solvation. The particle becomes encased in a sheath of solvent which acts as a further barrier thus preventing the particles from coalescing to form larger aggregates.

The removal of surface charge would be sufficient to bring about coagulation of lyophobic sols. In lyophilic sols, on the other hand, charge removal may decrease the stability, but it does not necessarily lead to coagulation. For example, sols of proteins at their isoelectric point are uncharged. Usually, they do not coagulate at this condition because they are still protected by a layer of water molecules. However, as soon as this layer is removed by some means, coagulation takes place.

Protective colloids   Lyophilic colloids are much more stable than lyophobic colloids because they are extensively solvated. The greater stability of lyophilic colloids can be conveniently used to improve the stability of lyophobic colloids. The process, by which a lyophobic colloid is prevented from coagulation by the action of electrolytes due to the previous addition of some lyophilic colloid, is called protection. The lyophilic colloid used to achieve this is generally called as a protective colloid. Thus a gold sol is protected from coagulation by the addition of a little gelatin.

The protective action of the lyophilic colloid is due to the formation of a protective shell surrounding each particle of the lyophobic colloid through which the oppositely charged ions of added electrolytes cannot easily penetrate so as to neutralize the charge. As a result, the lyophobic sols acquire greater resistance to coagulation.

The lyophilic colloids differ in their protective power. Zsigmondy introduced the term **gold number** to measure the protective powers of different colloids.

The gold number is defined as the weight (in mg) of a protective colloid which prevents the coagulation of 10 ml of a given gold sol on adding 1 ml of 10% sodium chloride solution.

Thus, the smaller the gold number, the greater the protective power of the lyophilic colloid. The gold number of some protective colloids are given in Table 5.3.

Table 5.3   Gold number of protective colloids

Protective colloid	Gold number
Gelatin	0.005–0.01
Casein	0.01–0.02
Gum arabic	0.15–0.25
Egg albumin	0.08
Starch	25.0

Coagulation of Sols

The colloidal particles are stable because of the existence of electric charge on them. Due to electrical repulsion, the particles do not come closer to each other and coalesce. If by any means the charge is neutralized, the particles can aggregate and coagulate immediately. This phenomenon by means of which the colloidal sol changes into a suspended state is known as **coagulation** or flocculation of colloids.

For coagulation of lyophobic sols, the charge on the particles must be removed, while for lyophilic, removal of both the charge and the adhering layer of solvent is necessary. Coagulation of sols is generally brought about by any one of the following methods.

By persistent dialysis   The traces of electrolytes are essential for the stability of sols. If the sols are subjected to prolonged dialysis, these traces of electrolytes also pass out through the dialyser and colloids become unstable.

By electrophoresis   The coagulation of a sol may be accomplished by electrophoresis, i.e., passing electric current through the sol. The charged particles on coming in contact with an electrode of opposite sign get discharged and coagulated.

By mutual action of sols   When two sols carrying opposite charges are mixed together in suitable proportions, mutual coagulation occurs, for example, coagulation of positively charged ferric hydroxide sol by negatively charged arsenic sulphide sol. The charges on both the sols are mutually neutralized and hence, both get coagulated.

By boiling and freezing   Coagulation can occasionally be effected by boiling or freezing. The process of boiling reduces the amount of adsorbed electrolytes responsible for stability. Freezing on the other hand results in the removal of medium.

By the action of electrolytes   Though traces of electrolytes are essential for the stability of sols, the presence of large amounts causes the coagulation. The reason is that colloidal particles take up oppositely charged ions from electrolytes with the result the charges on the sol particle are neutralized and coagulation takes place. The ion carrying the opposite charge is called the "flocculating ion". For example, arsenic sulphide sol or ferric hydroxide sol is precipitated by adding $BaCl_2$ or $Al_2(SO_4)_3$ solution. In this case, the negatively charged sol of arsenic sulphide takes up the barium or aluminium ions, while the positively charged sol of ferric hydroxide takes up chloride or sulphate ions. This causes neutralization of charge and hence coagulation of the sols.

The amount of electrolyte required to coagulate a fixed amount of sol depends upon the valency of the flocculating ion. Thus, "the coagulating power of an electrolyte depends predominantly on the valency of the flocculating ion, and higher the valency, greater is its coagulating power." This statement is known as **Hardy–Schulze rule**.

It has been observed that in the coagulation of arsenic sulphide sol, the flocculating power decreases in the order of $Al^{3+} > Ba^{2+} > Na^+$. Similarly in the coagulation of ferric hydroxide sol, the coagulating power decreases in the order $[Fe(CN)_6]^{4-} > PO_4^{3-} > SO_4^{2-} > Cl^-$.

The minimum concentration of an electrolyte required to cause flocculation of a sol is called its flocculation value. This is generally expressed in millimoles per litre. The reciprocal of flocculation value is regarded as the coagulating power. The flocculation value of some electrolytes for arsenic sulphide and ferric hydroxide sol are given in Table 5.4.

Table 5.4   Flocculation values of some electrolytes for As_2S_3 and $Fe(OH)_3$ sols

Electrolyte	Flocculation value (milli moles per litre)
Electrolytes for As_2S_3 sol	
NaCl	52
KCl	52
$BaCl_2$	0.69
$MgSO_4$	0.72
$AlCl_3$	0.093
Electrolytes for $Fe(OH)_3$ sol	
KNO_3	132
K_2CrO_4	0.315
K_2SO_4	0.210
$K_3[Fe(CN)_6]$	0.096
$K_4[Fe(CN)_6]$	0.085

EMULSIONS

An emulsion is a colloidal system in which the dispersed phase and dispersion medium are normally the liquids.

The common example of emulsion is milk in which the particles of liquid fat are dispersed in water. Cod-liver oil is another example in which particles of water are dispersed in oil. The particles of dispersed phase in emulsion are generally bigger than those in sols and are sometimes visible under the microscope.

Emulsions are unstable. However, they may be stabilized by the addition of emulsifying agents, like gelatin, soap, gum arabic, etc.

Types of Emulsions

Emulsions essentially belong to two types.

1. ***Water-in-oil emulsion (W/O type)***   In this type of emulsion, water is the dispersed phase and oil is the dispersion medium. The typical examples are butter, ice cream, etc.

2. ***Oil-in-water emulsion (O/W type)***   In such emulsion, oil forms the dispersed phase and water forms the dispersion medium. The typical examples are milk, vanishing cream, etc.

Preparation of Emulsion

The following methods are used in the preparation of emulsions.

Agent-in-oil method   In this method, the emulsifying agent is dissolved in oil, and water is added to it dropwise with constant stirring. This gives a W/O type emulsion.

Agent-in-water method   The emulsifying agent is dissolved in water, and the oil is added dropwise by continuous stirring. Thus an O/W type emulsion is formed.

Nascent soap method   This method is suitable in the case, where soap is being used as an emulsifying agent. In this method, the fatty acid part of the emulsifying agent is dissolved in oil and its alkali part in water, and then the two are mixed with constant stirring. The soap thus formed at the interface of oil and water gives a stable emulsion.

The type of emulsion formed depends on the phase-volume ratio, i.e., the relative volumes of oil and water in the emulsion.

Identification of Type of Emulsion

The water-in-oil or oil-in-water type emulsions can be identified by a number of ways.

Soluble dyestuff method   In this method, an oil-soluble dye is selected and added to the emulsion under examination. Since the dye is soluble in oil, then the whole emulsion becomes deeply coloured if it is of water-in-oil type. If no colour is developed by adding such a dye, then the emulsion is of oil-in-water type. In general if the colour spreads throughout the whole emulsion, the phase in which the dye is soluble is the continuous phase.

Electrical conductivity method   It is a good method for identifying the type of an emulsion. The method involves the measurement of electrical conductivity of the emulsion. The conductivity is appreciable if water is the dispersion medium (O/W type) but negligible if oil is the dispersion medium (W/O type).

Dilution method   The method is based on the principle that the emulsion can readily be diluted by the liquid which forms the dispersion phase. A small quantity of the emulsion to be tested is placed on a microscopic slide and a drop of water is added, which is then mixed properly. If the water drop added is mixed uniformly, then the emulsion is of oil-in-water type.

Filter paper method   This method is based on the fact that one component of the emulsion (i.e., water) spreads over the filter paper. A drop of emulsion to be tested is placed over a filter paper. If the liquid spreads readily, leaving a spot at the centre then the emulsion is of oil-in-water type. If it does not spread, then it is of water-in-oil type.

General Properties of Emulsions

Most of the properties of emulsions are the same as the properties of sols, but they differ in certain properties.

Particle size   The particles of dispersed phase in emulsions are generally bigger than those in sols and are sometimes visible under the microscope.

Viscosity   The viscosity of an emulsion is largely affected by the viscosity of the medium. It has been pointed out that the viscosity of the medium is also affected by the concentration of the dispersed phase.

Electrical conductivity   The oil-in-water type emulsions exhibit high electrical conductivity, while water-in-oil type emulsions show very low (or negligible) conductivity. It has been found that the conductivity of emulsions increases with an increase in temperature. For example, in the case of 5% emulsion of petroleum in water, the increase in temperature from 25 to 90°C increases the conductivity (2 to 3 times of the original value).

Electrophoresis   Like sols, the particles of dispersed phase in emulsion are electrically charged (usually carry negative charge). The particles migrate to oppositely charged electrodes under the influence of an electric field.

Dilution   It has been found that the emulsions may be diluted with any amount of the dispersion medium. The dispersed liquid if mixed will form a separate layer.

Brownian movement   Like sols, the emulsions are found to show the phenomenon of Brownian movement. However, the emulsions with larger particle size sometimes may not show the Brownian movement.

Optical properties   The light-scattering properties of emulsions with different particle sizes have been studied by various investigators. In the case of certain type of emulsions, a weak Tyndall cone is observed, and in the case of fibrous particles, mostly those of organic molecular colloids, the scattering of light is very weak.

Stability of Emulsions

The emulsion formed by the mere shaking of hydrocarbon oil with excess of water (or vice versa) is highly unstable and soon separates into two distinct layers, when allowed to stand. In order to get stable emulsions of fairly high concentrations, it is necessary to add another substance known as an emulsifying agent or emulsifier in small quantity.

Emulsifying agents are generally long-chain compounds with polar groups such as soaps, sulphonic acids and alkyl sulphates. The function of an emulsifier is to lower the interfacial tension between water and oil and in this way it facilitates the mixing of the two liquids. The soap molecules get concentrated at the interface between water and oil in such a way that their polar end is oriented towards the water whereas the hydrocarbon chain (R—) towards the oil, as shown in Figure 5.10.

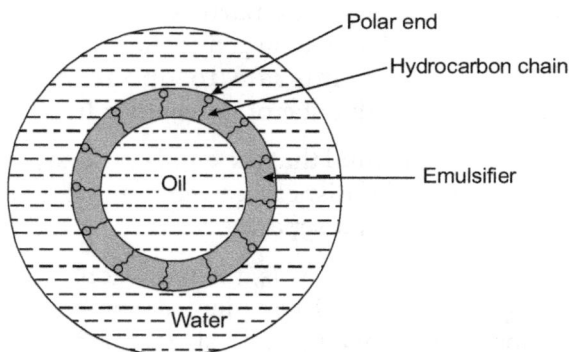

Figure 5.10   Structure of an emulsion particle

The sodium or potassium soaps produce oil-in-water-type emulsions, while calcium and aluminium soaps produce water-in-oil-type emulsions (provided the two liquids are present in proper proportions). There is no satisfactory explanation for this behaviour. Most of the lyophilic colloids like starch, glue, casein, gum and agar–agar are also used as emulsifiers. These substances form a protective layer around the particles of dispersed phase. Thus, in milk, casein is known to form a protective layer around fat globules dispersed in water.

GELS

Gel represents a liquid–solid system, i.e., the colloidal dispersion having a liquid dispersed in a solid of high concentration.

Many lyophilic sols (such as gelatin, agar–agar, gum arabic, etc.) and a few lyophobic sols (like silicic acid, ferric hydroxide sol, etc.), when coagulated under certain conditions, change into a semi-solid mass. Such a product is called a gel. The process of transformation of a sol into a gel is termed as gelation. The sols should be of sufficiently high concentration to facilitate the gelation process.

Preparation of Gels

Gels may be prepared by any one of the following methods.

By cooling of sols The gels of some sols like gelatin and agar–agar, are prepared by cooling their sols of moderate concentrations prepared in hot water. When cooled, the hydrated particles agglomerate together to form larger aggregates which ultimately form a semi-solid mass entrapping the entire liquid within itself.

The gelation of this type depends upon the temperature and time of gelation, viscosity of the medium, and the minimum concentration of the sol.

By double decomposition The hydrophobic gels (such as silicic acid and aluminium hydroxide gels) are formed by the process of double decomposition.

The silicic acid gel (or silica gel) is prepared by adding hydrochloric acid to an aqueous solution of sodium silicate. A highly hydrated precipitate of silicic acid is obtained which on standing sets into a gel.

The aluminium hydroxide gel (or alumina gel) is obtained by mixing solutions of sodium hydroxide and aluminium chloride of suitable concentrations. As a result, a highly hydrated precipitate of aluminium hydroxide is formed. On standing it changes into a gel.

By change of solvents   In the preparation of some of the gels, the gelation occurs by the change of solvent in which the sol is unstable. Thus in the preparation of calcium acetate gel, pure alcohol is added to an aqueous solution of calcium acetate, in this way whole of the calcium acetate goes into alcohol which then sets in the form of gel containing the liquid.

Elastic and Non-elastic Gels

Based on their properties, the gels can be divided into two types.

    i.   elastic gels

   ii.   non-elastic gels

These two types are distinguished mainly by their behaviour on dehydration and rehydration.

Elastic gels are reversible. When partially dehydrated, they change into a solid mass which changes back into the gel form, on simple addition of water followed by slight warming. Starch, agar-agar and gelatin gels are examples of elastic gels. Non-elastic gels, on the contrary, are irreversible. When dehydrated they become glassy or change into a powder which on addition of water and followed by warming does not change back into the original gel. Examples of non-elastic gels are silica, alumina and ferric oxide gel.

Properties of Gels

Elastic properties The gels show the elastic properties as in the case of a solid. They also exhibit the properties of flexibility, tensile strength, and compressibility.

Electrical conductivity It is found that there is no change in electrical conductivity when a sol is converted into a gel.

Optical properties The Tyndall effect is found to increase during the sol–gel transformation. It indicates that there is an increase in the size of particles during the change. Thus Tyndall effect is more pronounced in the case of gels.

Swelling or imbibition of gels It was observed that hydrophilic gels (such as gelatin, agar–agar and starch gels) have the property of absorbing definite amount of water or other liquids, causing the volume of the gel to increase. The swelling is very similar to imbibition which means the uptake of a liquid by porous material without any observed increase in volume.

Syneresis According to Graham, the syneresis is the spontaneous liberation of the liquid from a gel. Elastic and non-elastic gels both shrink when kept standing. The syneresis may be regarded as the reversal of swelling. It is observed in the case of starch and silica gel.

Thixotropy Some gels on shaking form the corresponding sol and on standing convert into the gel form. This phenomenon of reversible sol–gel transformation is known as thixotropy. Gelatin and silica gels show this property.

COLLOIDS IN PRINTING INDUSTRY

GELATIN

Gelatin is a protein produced by partial hydrolysis of collagen extracted from connective tissues of animals such as the domesticated bovines, porcines and equines.

On a commercial scale, gelatin is made from by-products of the meat and leather industry, mainly pork skins, pork and cattle bones, or split cattle hides. The type A gelatin is obtained from acid-treated raw materials and type B from alkali-treated raw materials. Acid pretreatment (type A gelatin) utilizes pig skin whereas alkaline treatment (type B gelatin) makes use of cattle hides and bones.

Gelatin can also be prepared by boiling certain cartilaginous cuts of meat or bones in water. Depending on the concentration, the resulting broth, when cooled, will naturally form a jelly or gel.

Gelatin is a translucent, brittle, colourless or slightly yellow, nearly tasteless and odourless solid substance. It melts when heated and solidifies when cooled again. Together with water it forms a semi-solid colloidal gel. Gelatin forms a solution of high viscosity in water, which sets to a gel on cooling and its chemical composition is, in many respects, closely similar to that of its parent collagen. If gelatin is put into contact with cold water, some of the material dissolves. Gelatin is soluble in hot water, glycerol and acetic acid. It is insoluble in alcohol, chloroform and other organic solvents. Gelatin gels exist over only a small temperature range, the upper limit being the melting point of the gel which depends on gelatin grade and concentration and the lower limit the ice point at which ice crystallizes out. Gelatin concentration and the temperature have important effects on viscosity. The higher they are, the higher their viscosity.

Applications

1. In photographic films, it is used to hold silver halide crystals in an emulsion in virtually all photographic films and photographic papers. Despite some efforts, no suitable substitutes with the stability and low cost of gelatin have been found.

2. In lithography, black-line sensitizer, which is prepared by mixing a solution of gelatin with citrate of iron and

ammonia, silver nitrate and water, is used to get black prints in offset lithography. Strip-film cement, which is prepared by mixing a solution of gelatin with acetic acid and water, is used in offset lithography.

3. As a surface sizing, it improves the smoothness of glossy printing papers or playing cards and maintains the wrinkles in crepe paper.

4. Gelatin is closely related to bone glue and is used as a binder in sandpaper.

GUM ARABIC

Gum arabic, a water-soluble natural gum (also called gum acacia) is obtained from several species of the acacia tree, especially *Acacia senegal* and *Acacia seyal*. Gum arabic is prepared from the exudates of stems and from the branches of *Acacia senegal* and *Acacia seyal* and produced naturally as large nodules during a process called gummosis to seal wounds in the bark of the tree.

Gum arabic is a complex mixture of polysaccharides and glycoproteins. It is a less-consistent material than other hydrocolloids. It is white or pale yellow in colour and odourless. With cold or hot water it gives a hydrophilic sol. It is insoluble in alcohol. It is available in thin flakes, powder or granules. It also reduces the surface tension of liquids.

Applications

1. In offset lithography, gum arabic is used in the preparation of deep-etch coating solutions. A solution of gum arabic containing ammonium dichromate $[(NH_4)_2Cr_2O_7]$ and ammonium hydroxide (NH_4OH) is used as deep-etch coating solution. This solution is used to coat the counter-etched lithoplates (Al, Zn, or stainless steel plates) to produce a light-sensitive film on the plate. Such plates are used to get images from a positive photographic transparency after exposure to an arc light. The non-image areas will be hardened while the image

areas will remain unhardened. After developing, such plates are used in offset printing.

2. In offset lithography, it is also an important ingredient in fountain solutions. The fountain solution usually contains gum arabic, phosphoric acid and other additives. When the developed lithoplate is treated with the above solution, a thin, invisible, adhering water-receptive gum film will be formed on the non-image areas. This will prevent the ink sticking to the non-image areas of the plate during printing process.

3. In photography, the historical photography process of gum bichromate photography uses gum arabic to permanently bind pigments on paper. Ammonium or potassium dichromate is mixed with gum arabic and pigment to create a photographic emulsion, sensitive to ultraviolet light.

4. It is used in the manufacture of ink. Historically, gum arabic was used to increase the viscosity of ink, or to make it flow well, to prevent it from feathering, and to suspend the colouring matter.

5. It is widely used as an adhesive on postage stamps and cigarette papers.

6. Printers employ it to stop oxidation of aluminum printing plates in the interval between processing of the plate and its use on a printing press.

7. Gum arabic is used as a binder for watercolour painting because it dissolves easily in water. Pigment of any colour is suspended within the gum arabic in varying amounts, resulting in watercolour paint.

EGG ALBUMIN

Egg albumin is a protein made from egg whites, the nutritive and protective gelatinous substance surrounding the yolk. It consists mainly of about 15% proteins dissolved in water. Unlike the egg yolk, it contains little fat. It is either a solid or liquid. It is white in colour.

Applications

1. In lithography, it is used in the preparation of deep-etch coating solution. A solution of egg albumin containing ammonium dichromate is used as deep-etch solution. This solution is coated on lithographic plates of Al, Zn, or stainless steel to get light-sensitive film on the plate. Such plates are used to get images for printing after exposure and developing. These plates are then used for printing.

2. It was used in the emulsion of traditional photographic paper.

CELLULOSE GUM

It is a synthetic product of cellulose and chemically known as sodium carboxy methyl cellulose.

It is a colourless, odourless, tasteless, non-toxic and hygroscopic solid. It is available in the form of powder or granules. It is readily dispersible in hot or cold water forming a hydrophilic sol. It forms a better hydrophilic sol compared to starch and gelatin. It remains insoluble in most of the organic solvents.

Application

In lithography, cellulose gum is used as a synthetic gum in dampening solutions. An aqueous solution containing phosphoric acid, magnesium nitrate, cellulose gum is one such dampening solution used in lithographic printing. Such a solution is used to desensitize the non-image areas on the lithoplates, so that these areas do not accept ink during printing process.

CASEIN GLUE

Casein glue is a protein derivative of skimmed milk. Casein is a protein present in milk and is the main ingredient of cheese. It is a phosphoprotein consisting of about 15 different amino acids. Casein glue is obtained by mixing casein with an alkaline solution of lime. It is one of the last commonly used natural glues—the use of animal glues derived from hides and bones being now all but obsolete. It is effective at both quite cold and hot temperatures, but is susceptible to moisture and fungal attack. It has very good adhesive property than any other animal glue.

Applications

1. In lithography, an aqueous solution of casein containing ammonium dichromate and ammonium hydroxide is used as a deep-etch coating solution. This solution is applied on lithoplates to get light-sensitive film. Such plates are used in offset lithography.
2. It is used as an adhesive. Casein glue is one of the oldest wood-working glues and is still popularly used.

FISH GLUE

It is an animal protein extracted from fish. Fish glue is prepared by heating the skin or bones of fish in water. The purest form of fish glue, made from the membrane of the air bladder (swim bladder) of certain species of fish such as the sturgeon, is also called isinglass. Isinglass can be produced from various species of fish using diverse manufacturing processes. Depending on the manufacture, the purity of isinglass may vary. Historic sources do not always specify which part of the fish was used to make the glue. It is flexible and having good adhesive property. But it shows poor resistance towards moisture and fungus.

Applications

1. Special high-grade fish glue, called **photoengraving glue,** is used in the graphic art industry to make blueprint papers, tracing cloths, and letterpress printing plates.
2. As we can see from ancient and medieval records, fish glue was both a common and important adhesive for many special applications; adapted by artists, it was used from

the time of ancient Egypt to twentieth-century France, in painting media, in coatings and grounds, in the gilding of illuminated manuscripts, and in pastel fixatives.

REVIEW QUESTIONS

1. What are colloidal dispersions? Discuss the types of colloidal dispersions.

2. Explain the following terms.
   i. Colloidal state
   ii. Dispersed phase
   iii. Dispersion medium

3. In what ways do colloidal dispersions differ from true solutions? In what way are they similar?

4. Explain the meaning of the following terms:
   i. Emulsion
   ii. Foam
   iii. Solid foam
   iv. Liquid aerosol
   v. Gels

5. What are sols? Distinguish between a lyophilic and lyophobic sol.

6. Give the preparation of the following:
   i. Arsenic sulphide sol
   ii. Sulphur sol
   iii. Gold sol
   iv. Ferric hydroxide sol

7. Describe origin of charges on colloidal particles.

8. Explain the electrical properties of colloids.

9. Explain the following methods of preparation of sols.
   i. By mechanical dispersion
   ii. By double decomposition

    iii. By oxidation

    iv. By change of solvent

10. Write short notes on:

    i. Electrophoresis

    ii. Brownian movement

    iii. Electro-osmosis

11. Why are colloidal particles electrically charged? How is the sign of electric charge on the colloidal particles determined experimentally?

12. What are protective colloids? How can a lyophilic colloid stabilize a lyophobic colloid?

13. Describe the preparation of a metal sol by Bredig's arc method.

14. Write short notes on:

    i. Peptization

    ii. Coagulation of sols

15. Explain the Hardy–Schulze rule.

16. Discuss the various factors determining the stability of colloidal dispersions.

17. Explain the kinetic properties of sols.

18. What do you mean by purification of sols? Discuss the various methods employed for the purification of sols.

19. Explain the following terms.

    i. Gold number

    ii. Flocculation value

20. What are emulsions? Name the types of emulsions.

21. Explain the methods of preparation of emulsions.

22. Describe the methods used in the identification of types of emulsions.

23. What are gels? Describe the various methods for the preparation of gels.

24. Which of the following electrolytes is most effective in coagulating arsenic sulphide sol: $MgSO_4$, NaCl, $AlCl_3$? Justify your answer.

25. Gold number for gelatin, haemoglobin and sodium oleate is 0.005, 0.05, and 0.7 respectively. Which of these has the greatest protective power and why?

26. Discuss the functions of an emulsifier.

27. Distinguish between elastic and non-elastic gels.

28. Discuss the applications of the following colloids in the printing industry.
    i. Gelatin
    ii. Gum arabic
    iii. Egg albumin
    iv. Casein glue
    v. Fish glue

29. "A colloidal dispersion is not precipitated on addition of an electrolyte in the presence of gelatin." Explain why.

ADDITIONAL READING

Holmberg, K., Shah, D.O. and Schwuger, M.J. (2002). *Hand Book of Applied Surface and Colloid Chemistry*. John Wiley and Sons, New York.

Hunter, R.J. (2003). *Foundation of Colloid Science*, 2nd edn. Oxford University Press, New York.

Maron, S.H. and Prutton, C.F. (1972). *Principles of Physical Chemistry*, 4th edn. Oxford and IBH Publishing Co. Pvt. Ltd., New Delhi.

6 DYES AND PIGMENTS

INTRODUCTION

A dye can generally be described as a coloured substance that has an affinity to the substrate to which it is being applied. The dye is generally applied in an aqueous solution, and may require a mordant to improve the fastness on the fibre.

Both dyes and pigments appear to be coloured because they absorb some wavelengths of light preferentially. In contrast to a dye, a pigment generally is insoluble, and has no affinity for the substrate. Some dyes can be precipitated with an inert salt to produce a lake pigment.

COLOUR AND CHEMICAL CONSTITUTION

When white light (electromagnetic radiation with wavelength in the range of 400–800 nm or 4000–8000 Å) falls on a substance, it may be of any of the following.

i. totally reflected—the substance in this case will appear white
ii. totally absorbed—the substance will appear black
iii. partially absorbed and partially reflected—the substance will have the colour of light reflected.

The visible colour is complementary to the colour absorbed. Thus a substance may appear blue either because it absorbs all other wavelengths except the wavelength corresponding to blue light (435–480 nm), or because it absorbs yellow light (580–595 nm) which is complementary to blue. The list of complementary colours for various wavelengths is given in Table 6.1.

Table 6.1   Relationship of absorbed and visible colour

Wavelength absorbed (nm)	Colour absorbed	Complementary colour
400–435	Violet	Yellow-green
435–480	Blue	Yellow
480–490	Green-blue	Orange
490–500	Blue-green	Red
500–560	Green	Purple (magenta)
560–580	Yellow-green	Violet
580–595	Yellow	Blue
595–605	Orange	Green-blue
605–750	Red	Blue-green (cyan)

Otto Witt (1876) was the first one to formulate a theory to explain the colour of organic compounds and the theory is known as "chromophore–auxochrome theory". He pointed out that

1. The colour of an organic compound is due to the presence of certain multiple-bonded groups called chromophores (Greek: *chromo* = colour, *pherein* = to bear). Some of the important chromophores are given below.

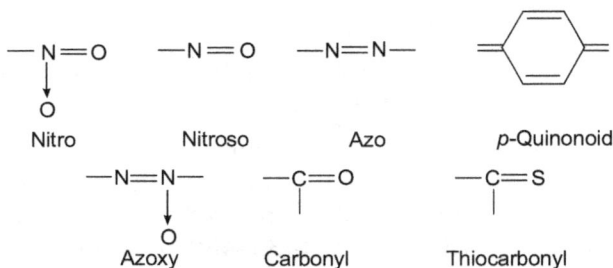

Nitro   Nitroso   Azo   p-Quinonoid

Azoxy   Carbonyl   Thiocarbonyl

Compounds containing the chromophoric groups are named as **chromogens**.

2. The intensity of colour increases with an increase in the number of chromophores in a chromogen. This effect is particularly marked if the chromophores are conjugated with an extensive system of alternate single and double bonds as exists in aromatic rings.

3. Certain groups, which are not producing colour by themselves, when present in a chromogen, intensify the colour. Such colour-assisting groups are termed as auxochromes (Greek: *auxein* = to increase). They may be either an acidic or a basic functional group.

Acidic: —OH (hydroxy); —SO₃H (sulphonic acid); — COOH (carboxylic acid)

Basic: —NH₂ (amino); — NHR (alkyl amino);   — NR₂ (dialkyl amino)

Nitrobenzene is a pale yellow substance but when the auxochrome —OH is present in ortho- or para position, the product becomes deep yellow. The auxochrome —OH deepens the colour by extending the conjugated system between the chromophore and the auxochrome due to resonance.

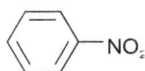

Nitrobenzene (pale yellow)   *p*-Nitrophenol (deep yellow)

In general, it is the chromophores that produce the colour and the auxochromes that deepen the colour. Groups that shift the absorption peak to the longer wavelength region are called the bathochromic groups and the effect as **bathochromic shift**. Those which cause the opposite effect are termed as hypsochromic groups and the effect as **hypsochromic shift**. Generally, auxochromes are found to be bathochromic, the effect being increased by the replacement of H in the NH₂ group with alkyl or aryl groups. On the other hand, the replacement of H in OH or NH₂ by an acetyl group generally has a hypsochromic effect.

Recently, the chromophore–auxochrome theory has been reinterpreted and revised in terms of electronic theory. According to this theory, chromophore is defined as any isolated functional group that shows absorption in the ultraviolet or visible region (200–800 mμ). Similarly, auxochrome is defined as the group that increases the absorption of a chromophore. The groups with multiple bond or π electrons (chromophores), or groups with non-bonding electrons (usually auxochromes) shift the absorption band to longer wavelength. If the number of these groups present is sufficient to shift the absorption to visible region, the compound will appear coloured.

DYES

Dyes are natural or synthetic coloured organic compounds that impart colour to the substrate (such as textile materials) to which they are being applied. After dyeing, the dye must be fast to light (not to be bleached by light) and chemically stable so as to resist the action of soap detergents, water, acids, alkali or dry-cleaning solvents.

The first human-made (synthetic) organic dye, mauveine, was discovered by William Henry Perkin in 1856. Many thousands of synthetic dyes have since been prepared. Synthetic dyes quickly replaced the traditional natural dyes. They cost less, offer a vast range of new colours, and impart better properties upon the dyed materials.

NOMENCLATURE OF DYES

Generally, the chemical names of dyes are very complicated and therefore the trade names (or the name given by the manufacturers) are commonly used. These trade names often include an alphabetical letter indicating the shade of the dye, for instance, B for blue, O for orange, R for red, Y or G for yellow (German: *gelle*), etc. Sometimes these letters refer to specific properties of the dye, for instance, F for fast to light,

WF for fast to washing (German: *waschechi*), D for direct dye, S for acid dye (German: *sauer*), etc.

The society of dyers and colourists has compiled a list of names and colours of various dyes, known as colour index. In this index, each dye has been assigned its individual colour number (CI number).

CLASSIFICATION OF DYES

Dyes may be classified on the basis of their chemical structure or according to the method of their application to the fibre. The former classification is of more interest to chemists, who desire to know what makes the molecule the colour it is and the latter classification is of more interest to dyers, who need to know which dye is appropriate to the material they need to dye, and the resultant colour.

Classification Based on Structure

Dyes may be classified according to the types of chromophores present in their structures. The classification includes the following main types.

Nitro and nitroso dyes   They are the earliest known synthetic dyes which contain nitro ($-NO_2$) or nitroso ($-NO$) groups as chromophores in their molecules. These dyes may also contain hydroxyl ($-OH$) or amino ($-NH_2$) or sulphonic acid ($-SO_3H$) groups as auxochromes. Some examples:

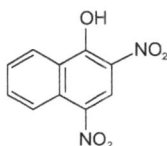

Martius yellow   Naphthol yellow S   Naphthol green Y

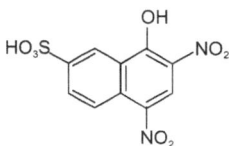

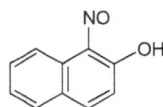

Martius yellow (2, 4-dinitro-1-naphthol) is prepared by nitrating 1-naphthol 2, 4-disulphonic acid. It is used as a direct dye for wool and silk.

Naphthol yellow S is obtained by nitrating 1-naphthol 2, 4, 7-trisulphonic acid. It is used as an acid dye for wool and silk.

Naphthol green Y is obtained by the reaction of naphthols with nitrous acid. It is used as a mordant dye for wool.

Azo dyes   Azo dyes form the largest and most important group of synthetic dyes. They are characterized by the presence of azo $(-N=N-)$ group as chromophore. The common auxochromes present in azo dyes are $-NH_2$, $-NR_2$, $-OH$, $-SO_3H$, etc. They may be further classified as monoazo, diazo, triazo, etc. depending on the number of azo groups present. They are highly coloured and can be prepared by diazotizing a primary amine and by subsequent coupling with suitable aromatic amines or phenols.

The reaction of primary aromatic amines with nitrous acid (HNO_2, an unstable reagent, generated by the action of mineral acid like HCl on sodium nitrite, $NaNO_2$) gives the product known as diazonium salt and the reaction is known as **diazotization**.

By varying the substituents present in both the diazonium salt and the coupling compound, a series of azo dyes can be produced with almost any colour. Some of the examples are discussed in the following section.

Aniline yellow (p-amino azobenzene)   It is a basic azo dye prepared by coupling benzene diazonium chloride with aniline.

Benzene diazonium chloride   Aniline   Aniline yellow

It is not used directly as a dye because of its acid-sensitive nature. However, it is used as an intermediate in the preparation of other dyes.

Methyl orange   It is an important acidic azo dye prepared by coupling diazotized sulphanilic acid with dimethylamine.

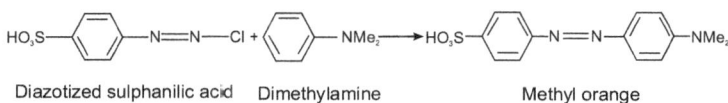

HO₃S—⟨ ⟩—N=N—Cl + ⟨ ⟩—NMe₂ ⟶ HO₃S—⟨ ⟩—N=N—⟨ ⟩—NMe₂

Diazotized sulphanilic acid   Dimethylamine   Methyl orange

It is an acid dye used for wool and silk. It is used as an indicator in acid–base titrations because it gives an orange-yellow colour in alkaline solutions and a red colour in acidic solutions.

Bismarck brown   It is a basic azo dye obtained by coupling tetrazotized *m*-diaminobenzene (*m*-phenylene diamine) with *m*-diamino benzene (2 moles).

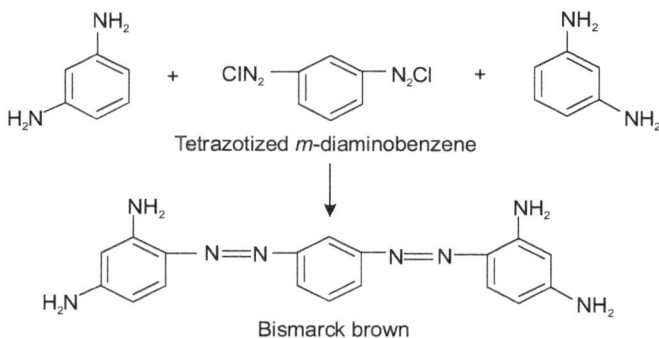

Tetrazotized *m*-diaminobenzene

Bismarck brown

It is a brown-coloured dye used in boot and wood polishing. It dyes wool and mordanted cotton.

Diphenyl methane dyes   They are characterized by a basic diphenyl methane skeleton. An example for this type of dye is given below.

Auramine G   It is a greenish yellow basic dye, obtained by the condensation reaction of formaldehyde with *N*-monomethyl toluidine.

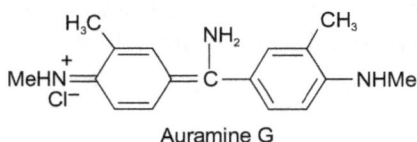

Auramine G

Triphenyl methane dyes They have the basic triphenyl methane skeleton. The following are some examples for this type of dyes.

Malachite green It is prepared by the condensation reaction of benzaldehyde (1 mole) with N,N-dimethyl aniline (2 moles) in the presence of a dehydrating agent like concentrated sulphuric acid at 100°C. The leuco-base so formed is oxidized with lead dioxide followed by treatment with excess hydrochloric acid yielding malachite green.

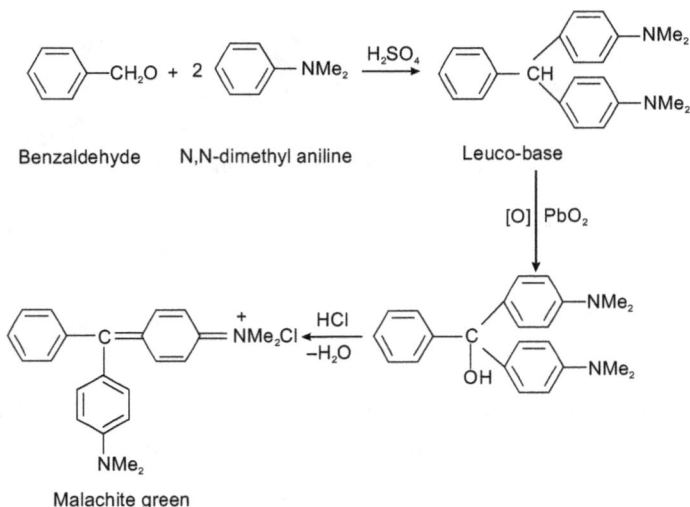

Benzaldehyde   N,N-dimethyl aniline   Leuco-base

Malachite green

It has a deep blue colour and not fast to light. It dyes wool and silk directly, and cotton mordanted with tannin.

Pararosaniline It is prepared by the oxidation of *p*-toluidine (1 mole) and aniline (2 moles) with nitrobenzene followed by treatment with hydrochloric acid.

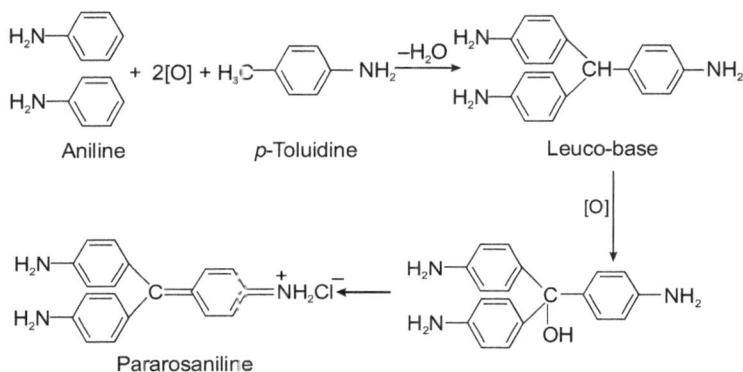

Aniline   *p*-Toluidine   Leuco-base

Pararosaniline

It dyes silk and wool directly, and cotton mordanted with tannin.

Rosaniline   It is obtained by the oxidation of an equimolar mixture of aniline, *o*-toluidine and *p*-toluidine with nitrobenzene in the presence of iron fillings and by treatment with hydrochloric acid.

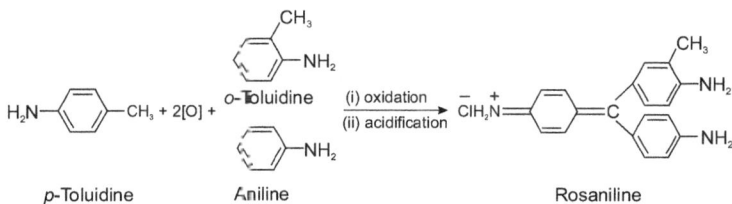

p-Toluidine   Aniline   Rosaniline

It dissolves in water to give a deep red solution. The solution is decolorized by sulphur dioxide and is used as Schiff's reagent for the detection of aldehydes. It is used as a direct dye for wool and silk. It dyes cotton after mordanting with tannin.

Anthraquinone dyes   Anthraquinone dyes are derived from anthracene. Alizarin (1, 2-dihydroxy anthraquinone), the most important dye, belongs to this class. It is manufactured by sulphonating anthraquinone with fuming sulphuric acid (oleum) at 140°C. The sodium salt of the product, anthraquinone-2-sulphonic acid on fusion with sodium hydroxide in the presence of sodium or potassium perchlorate at 200°C under pressure, gives alizarin.

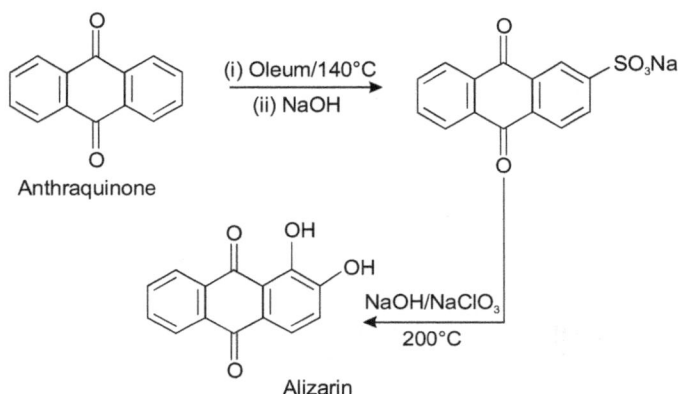

Anthraquinone

(i) Oleum/140°C
(ii) NaOH

SO₃Na

Alizarin

NaOH/NaClO₃
200°C

Alizarin forms ruby red crystals, which dissolve in alkalis to give a purple solution. It is the best known mordant dye. Aluminium and iron lakes are used in dying cotton and in printing inks.

Anthraquinone vat dyes are widely used in the textile industry. They have a brilliant colour with exceptional fastness. Indanthrone blue (Vat blue O) is an important vat dye.

Phthalocyanine dyes   They are the important types of synthetic organic dyes. Phthalocyanine dyes are very fast to light, heat, acids and alkalis. They are widely used in paints, printing inks, synthetic plastics and rubber materials.

Monastral Fast Blue was the first commercially produced phthalocyanine dye. It is a copper phthalocyanine dye, and is prepared by heating phthalic anhydride with urea and cuprous chloride (Cu_2Cl_2) in the presence of boric acid as a catalyst.

Classification Based on Application

Dyes are now classified according to how they are used in the dyeing process.

Acid dyes These are water-soluble anionic dyes that are applied to fibres such as silk, wool, nylon and modified acrylic fibres using neutral to acid dye baths. They are generally the sodium salts of sulphonic acids and nitrophenols. Attachment to the fibre is attributed, at least partly, to salt formation between anionic groups in the dyes and cationic groups in the fibre. Acid dyes are not substantive to cellulosic fibres. For example, Martius yellow (2, 4-dinitro-1-naphthol) and Naphthol yellow S are acid dyes.

Basic dyes They are water-soluble cationic dyes that are mainly applied to acrylic fibres, but find some use for wool and silk. Usually acetic acid is added to the dye bath to help the uptake of the dye on to the fibre. Basic dyes are also used in the colouration of paper. Dyes belonging to this type are malachite green, rosaniline, crystal violet, etc.

Direct dyes A direct dye can be applied to fabric by direct immersion in a hot aqueous solution of the dye. Direct or substantive dyeing is normally carried out in a neutral or slightly alkaline dye bath, at or near boiling point, with the addition of either sodium chloride or sodium sulphate. Direct dyes are used on cotton, paper, leather, wool, silk and nylon. They are also used as pH indicators and as biological stains. Martius yellow is a typical direct dye.

Mordant dyes They require a mordant, which improves the fastness of the dye against water, light and perspiration. A mordant is that which provides a chemical link or bridge between the dye and the fibre. The choice of mordant is very important as different mordants can change the final colour significantly. Most natural dyes are mordant dyes and there is therefore a large literature base describing dyeing techniques. The most important mordant dyes are the synthetic mordant dyes, or chrome dyes, used for wool; these comprise some 30% of dyes used for wool, and are especially useful for black and navy shades. The mordant, potassium dichromate, is applied as an after-treatment. It is important to note that many mordants, particularly those under the hard metal category, can be hazardous to health and extreme care must be taken in using them. Alizarin is a typical mordant dye, which relies on a metal complex of trivalent chromium, to bond the dye molecule to the side groups of the fibre molecule.

In the paper industry, alum (potassium aluminium sulphate) is used to fix rosin size to paper fibres in order to improve their water- and ink-controlling properties, and also fix dyestuffs in paper.

Vat dyes   They are essentially insoluble in water and incapable of dyeing fibres directly. However, on reduction with alkaline sodium hyposulphite, they are converted into water-soluble colourless compounds (leuco base). The leuco-form is readily absorbed by the textile fibres. After dyeing with the leuco base, the textile material is exposed to air (oxygen). Thus the leuco form is oxidized to the original insoluble dye. Indigo is a typical vat dye.

Reactive dyes   These dyes have the ability to react chemically with the fibre substrate to produce permanent bond. They usually react with the –OH groups of cellulosic fibres or with the $-NH_2$ groups of proteinous fibres to form ester or amide linkages. The covalent bonds that attach reactive dye to natural fibres make it among the most permanent of dyes. "Cold" reactive dyes (such as Procion MX, Cibacron F, and Drimarene K) are very easy-to-use because the dye can be applied at room temperature. Reactive dyes are by far the best choice for dyeing cotton and other cellulose fibres at home or in the art studio. The vegetable dye henna (a hair dye) reacts with keratin in hair protein, permanently staining the hair until it grows out.

Disperse dyes   They were originally developed for the dyeing of cellulose acetate, and are substantially water-insoluble. The dyes are finely ground in the presence of a dispersing agent and then sold as a paste, or spray-dried and sold as a powder. They can also be used to dye nylon, cellulose triacetate, polyester and acrylic fibres. In some cases, a dyeing temperature of $130°C$ is required, and a pressurized dye bath is used. The very fine particle size gives a large surface area that aids dissolution to allow uptake by the fibre. The dyeing rate can be significantly influenced by the choice of dispersing agent used during the grinding. Generally these dyes are not fast to light and washing. 1-Amino-4-hydroxy anthraquinone (Disperse Red 15) is an important dye of this type.

Sulphur dyes These are sulphur-containing dyes that are applied to fibres (usually cotton) from aqueous sodium sulphide solution because they are insoluble in water. The soluble reduced form of the dye on the fabric is converted into its original insoluble form by aerial or chemical oxidation. The initial bath imparts a yellow colour. This is oxidized in place to produce the dark black we are familiar with in socks and the indigo blue of the common blue jeans.

PIGMENTS

Pigments have probably been used as colourants for even longer than dyes. The prehistoric cave paintings show the use of red, brown and yellow mineral earths. A pigment is a material that changes the colour of light it reflects as the result of selective colour absorption. This physical process differs from fluorescence, phosphorescence, and other forms of luminescence, in which the material itself emits light. Many materials selectively absorb certain wavelengths of light. Materials that humans have chosen and developed for use as pigments usually have special properties that make them ideal for colouring other materials. A pigment must have a high tinting strength relative to the materials it colours. It must be stable in solid form at ambient temperatures. For industrial applications as well as in the arts, permanence and stability are desirable properties. Pigments that are not permanent are called fugitive. Fugitive pigments fade over time or on exposure to light, while some eventually blacken.

Pigments are coloured or colourless finely divided organic or inorganic solids which are usually insoluble in the vehicle or medium in which they are incorporated. They are usually incorporated by dispersion in a variety of systems and retain their crystal and particulate nature throughout the pigmentation process. In contrast to dyes, whose colouristic properties are almost exclusively defined by their chemical structure, the properties of pigments are also defined by the physical characteristics of its particles. The physical factors that can modify the inherent colour of pigments are particle size, crystal morphology and shape. In general, pigments provide high degree of fastness and opacity.

Pigments are used for colouring paint, ink, plastic, fabric, cosmetics, food and other materials. Most pigments used in manufacturing and the visual arts are dry colourants, usually ground into a fine powder. This powder is added to a vehicle (or matrix), a relatively neutral or colourless material that acts as a binder.

PROPERTIES OF PIGMENTS

Selection of a pigment for a particular application is determined by cost, and by the physical properties and attributes of the pigment itself. The properties of a pigment depend not only on its chemical structure but also on the physical attributes of the pigment. The main essential properties of a pigment are its strength, brightness or saturation, fastness, dispersibility, and hiding power and transparency.

Strength   The inherent strength of a pigment depends on its light-absorbing characteristics which are related to its molecular and crystalline structure. In addition, strength is a function of particle size or surface area. The ability of a pigment to absorb light increases with decreasing particle size or increasing surface area, until the particles become entirely translucent or transparent to incident light. Pigment strength in a vehicle also depends on the typical character of other components in a pigmented system insofar as they absorb or scatter light. Strength comparisons are usually made with series of samples featuring varying amounts of a pigment incorporated in a vehicle with a corresponding series in the same vehicle containing a reference pigment. Instrumental comparisons are commonly practiced.

Brightness or saturation   The saturation of a coloured pigment is a measure of its brightness or cleanliness as apposed to dullness of hue. Generally, if a pigment absorbs light over a wide range of wavelength, i.e., shows broad absorption bands, or contains more than one chromophore, the pigment is likely to be duller than a pigment with sharp absorption bands due to a single chromophore. Because pigments are frequently used in combinations or blends, the brightness is determined by the selective absorption of the individual pigments and this significantly affects the brightness of the reflected colour.

Fastness   Fastness describes the characteristics of a pigment in terms of its colour stability in a pigmented system upon exposure to light, weather, heat, solvents, or various chemical agents. Ideally, a pigment should be insoluble, and chemically and photochemically inert. Only a few organic pigments approach such perfection.

Light fastness   Many pigments when exposed to high-intensity light such as direct sunlight or UV lamp, can get darker, change their shade or lose the colour saturation. This tendency must be minimized where the printed product is likely to be exposed to UV light. The colour and its saturation change mainly for organic pigments. Inorganic pigments, particularly those containing ions that can exist in several oxidation states (such as Pb, Cu, Cr, etc.), usually get darker. Some pigments are photoactive and can accelerate the decomposition of the organic matrix. Chalking caused by titanium dioxide, particularly its anatase form, is the best known example of a pigment accelerating the decomposition of the organic matrix.

Light fastness is measured by exposing pigmented film to an artificial or natural light for predetermined time.

Weathering   It is the ability of the coloured system (and not the pigment alone) to resist light and environmental conditions. The changes in colour and gloss are two main factors that are evaluated in weather tests.

Heat and chemical resistance   Heat stability is measured as a change in the hue of the coloured system and a degree of yellowing of the white system after exposure to a desired high temperature for a certain time. In determining the chemical resistance, colour changes of pigmented binder surfaces are measured after their exposure to various chemicals, such as water–sulphur dioxide or water–sodium chloride system. These systems imitate the environment to which the coloured articles could become exposed.

Dispersibility   The dispersibility of a pigment is measured by the effort required to develop the full tinctorial potential of a pigment in a vehicle system. Dispersibility differs from system to system depending on pigment–medium interaction and compatibility. The surface energy or wettability of the pigment should be compatible with the solvents and oils used in inks. The pigment wetting and dispersibility are dependent on the polarity relationship between pigment and dispersant (polar substances disperse in polar media whereas non-polar substances in non-polar media). It is found that the cube-shaped form of β-copper phthalocyanine with high surface polarity disperses more readily in polar solvents. Hence, it is often used in alcohol-based flexographic and gravure inks. On the other hand, the rod-shaped form having lower surface polarity is most suitable for use with non-polar media such as petroleum-based solvents. It is used in traditional litho inks.

Hiding power and transparency   Hiding power of a pigment is a function of its strength, that is, its absorption coefficients or light scattering coefficient, its particle size, and the relative refractive indices of the pigment and vehicle.

A pigment which absorbs much light increases hiding even when light scattering is insufficient, and the higher the refractive index, the greater is the hiding power of a pigment. In practical systems, hiding is determined by the ability of a pigment dispersion to completely hide or cover up a black and white checkered board.

To increase transparency, the light scattering must be minimized by particle size reduction. The smaller the particle size and the better the dispersion, the greater is the transparency.

Development of Synthetic Pigments

The industrial and scientific revolutions brought a huge expansion in the range of synthetic pigments (manufactured or refined from naturally occurring materials), which are widely used as colourants. Because of the expensiveness of Lapis Lazuli, much effort went into finding a less costly blue pigment. Prussian blue was the first synthetic pigment, discovered by accident in 1704. By the early 19th century, synthetic and metallic blue pigments had been added to the range of blues, including French Ultramarine, a synthetic form of Lapis Lazuli, and the various forms of Cobalt and Cerulean Blue. In the early 20th century, Phthalo Blue, an organic pigment with overwhelming tinting power, was synthesized.

Many manufacturers of paints, inks, textiles, plastics, and colours have voluntarily adopted the Colour Index International (CII) as a standard for identifying the pigments that they use in manufacturing particular colours. First published in 1925, and now published jointly on the web by the Society of Dyers and Colourists (United Kingdom) and the American Association of Textile Chemists and Colourists (USA), this index is recognized internationally as the authoritative reference on colourants. It encompasses more than 27,000 products under more than 13,000 generic colour index names.

In the CII scheme, each pigment has a generic index number that identifies it chemically, regardless of proprietary and historic names. For example, phthalo blue has been known by a variety of generic and proprietary names since its discovery in the 1930s. In much of Europe, phthalocyanine blue is better known as Helio Blue, or by a proprietary name such as Winsor Blue. An American paint manufacturer, Grumbacher, registered an alternate name 'Thalo Blue' as a trademark. Colour Index International resolves all these conflicting historic, generic, and proprietary names so that manufacturers and consumers can identify the pigment (or dye) used in a particular colour product. In the CII, all phthalo blue pigments are designated by a generic colour index number as either PB15 or PB36, short for pigment blue 15 and pigment blue 36. The two forms of phthalo blue, PB15 and PB36, reflect slight variations in molecular structure that produce a slightly more greenish or reddish blue.

TYPES OF PIGMENTS

Pigments can be conveniently divided into four types.

1. carbon black,
2. organic pigments,
3. white pigments and
4. coloured inorganic pigments.

Carbon Black

Carbon black is one of the oldest pigments known. It was used by prehistoric peoples for painting picture on cave walls, and commercially produced by the Chinese as early as 3000 BC.

It consists of pure elemental carbon (graphite), and it appears black because it reflects almost no light in the visible part of the spectrum. A variety of processes are used to produce carbon black, and the properties of the pigment vary somewhat depending on the manufacturing process employed.

Furnace blacks and channel blacks are the common blacks which are widely used in printing inks. Furnace blacks are produced by burning either oil or natural gas in a furnace with a limited supply of oxygen. They produce particles in the size range 30–150 μm and are separated with electric precipitators or cyclone collectors. Gas furnace blacks have the largest particle size and consequently produce matt and tinctorially weak inks which affect the opacity of ink films. Channel blacks are also made from natural gas but the particles are collected on moving iron channels. They produce the best quality, long-flowing blacks with particles in the size range, 20–30 μm.

Carbon black is the most important pigment used in black inks and toners. It is relatively cheap and possesses remarkable properties including good colour strength, and excellent resistance to light, heat, moisture and most chemicals. Carbon blacks have a much smaller particle size than any other pigment. Environmentally, carbon blacks are relatively stable, non-reactive and non-toxic.

Organic Pigments

Organic pigments are the largest group of pigments extensively used in printing inks.

Characteristics of organic pigments The important characteristic features of organic pigments are listed below.

- They are available in wide range of colours.
- They provide brighter shades and superior colour strength.
- They have a low specific gravity and small particle size that help to produce inks with good working properties.
- They have a soft texture and are free from problem of wear.

- They are easy to grind and produce inks with low abrasion characteristics.
- They are usually transparent pigments and produce transparent inks.
- The resistance of organic pigments to light, heat and chemicals, varies a great deal.

Some of the organic pigments may have poor resistance to soaps and oils, a tendency to bleed in solvents, a moderate stability to light and heat. Pigments that dissolve or bleed into water or dampening solution are unsatisfactory for lithographic inks. Some organic pigments sublime because of the heat generated in the delivery pile. When a pigment sublimes, it changes from solid to vapour. The vapour may transfer to the next sheet or even penetrate into several nearby sheets before it cools and resolidifies, producing a tint of the pigment. But such pigments are useful for the textile heat transfer printing process.

The chemistry of organic pigments   The chemistry of organic pigments is highly complicated. Most of the organic pigments are the derivatives of benzene, naphthalene, or anthracene. Pigment molecules also contain groups of atoms, called chromophoric groups, which are responsible for their colour. Some such chromophoric groups are $=C=NH-$, $=CH=N-$ and $-N=N-$.

Dyes are not usually suitable as colouring agents in printing inks, since they are soluble in common solvents. But dyes can be converted into pigments through several different chemical reactions. Basic dye pigments are prepared by the reaction of basic dyes with phosphomolybdic acid or phosphotungstic acid. When reacted with both, they are known as PMTA pigments. The name has been given because they are the complex salts of phospho-molybdic and tungstic acid. These pigments have strong brilliant colours and good light fastness. Pigments of this type include the rhodamines, Victoria blue, and the red, violet, blue and green PMTA pigments.

Azo pigments form the largest group of organic pigments. They contain the azo group, —N=N—, as chromophore. Generally, they are prepared by diazotizing an aromatic amine followed by coupling of the resulting diazonium salt. Azo pigments used in printing inks include Hansa yellow, diarylide yellows, lithol rubines, toluidine red, and red lake C.

Cyan, magenta and yellow pigments are extensively used in multicolour printing. The cyan pigment is copper phthalocyanine blue (CI Pigment Blue 15); Magenta is calcium 4B toner (Pigment Red 57:1) and yellow pigment is benzidine yellow (Pigment Yellow 13).

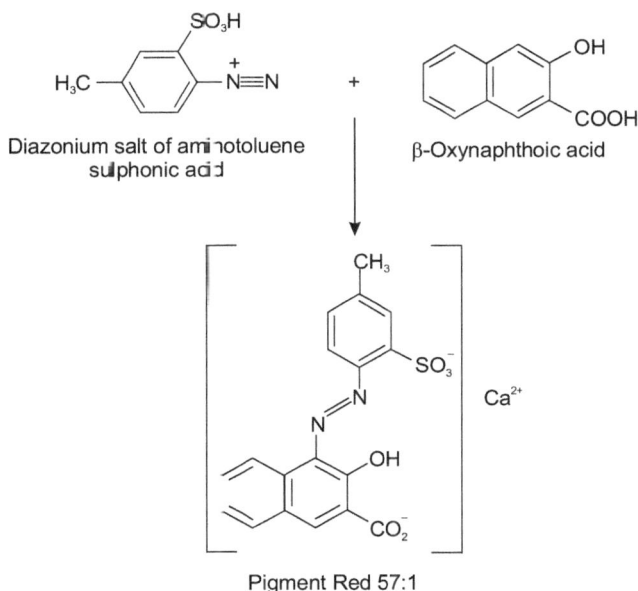

Diazonium salt of aminotoluene
sulphonic acid

β-Oxynaphthoic acid

Pigment Red 57:1

The yellow and magenta pigments are both azo compounds and thus contain the azo group (—N=N—). The chromophore in these systems is the —N=N— acceptor group coupled with the extended π-electron system of the benzenoid groups.

The magenta pigment, calcium 4B toner (Pigment Red 57:1) is prepared by diazotizing aminotoluene sulphonic acid (called 4B acid) and subsequent coupling with β-oxynaphthoic acid.

Benzidine yellow (Pigment yellow 13) is prepared by coupling the diazonium ion of 3, 3´-dichlorobenzidine with 2, 4-acetoacetoxylidide.

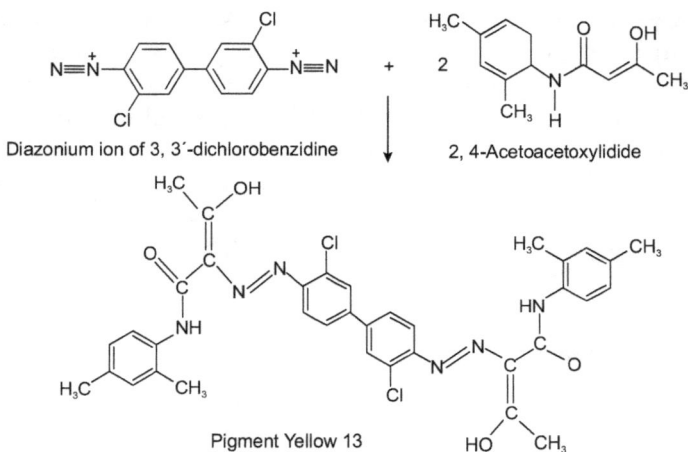

Diazonium ion of 3, 3´-dichlorobenzidine   2, 4-Acetoacetoxylidide

Pigment Yellow 13

Phthalocyanine pigments are made by heating phthalic anhydride with urea and cuprous chloride (Cu_2Cl_2). Such pigments are fast to light and alkalis.

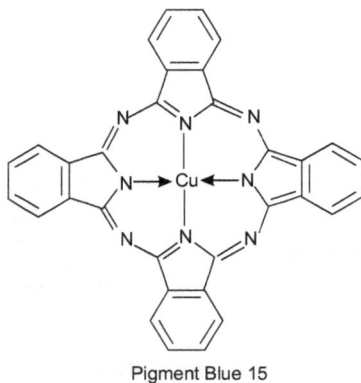

Pigment Blue 15

In the copper phthalocyanine molecule, the chromophore is the extended π-electron system. They resist bleeding since they are highly insoluble.

Flushed pigments Organic pigments can be obtained as flushed pigments. It is usually prepared as a dye and then converted to a pigment by precipitating it out of a water solution. The precipitate is then filtered, washed, and finally pressed to remove water content. The press cake obtained still contains some water, which can be removed by the process, called flushing. In the process, the press cake is mixed with a viscous varnish (such as linseed alkyd) in a large dough mixer. The varnish gradually displaces the absorbed water on the pigment surface, and the pigment slowly passes from the aqueous phase to the oil phase. The separated water is removed off. The traces of water in press cake can be removed by heating it under vacuum.

The advantages of flushed pigments are:

- They have high colour strength.
- They contain low moisture content.
- They can be readily mixed with other ink ingredients.

Some of the disadvantages of flushed pigments are

- Lack of flexibility in vehicle formation
- Low opacity
- Colour and body changes occur during manufacture and storage

Flushed pigments find greatest use where the drying of pigment results in the formation of hard and difficult-to-grind particles. In such cases, pigments are best prepared by the flushing process. But the flushing process is not suitable for certain heat-sensitive pigments, which must be air dried at low temperature to avoid colour changes.

Organic pigments are commonly flushed with alkyd or rosin ester varnishes. The use of flush pigments in quick set inks improves their fast setting properties.

White Pigments

They may be opaque or transparent.

- An opaque white pigment reflects light from their outer surface. They have the property of covering or hiding the background on which they are printed, e.g. titanium dioxide, zinc sulphide, lithopone, zinc oxide (in the decreasing order of opacity).

- Transparent white pigments do not reflect light at the surface. They allow the light to pass through the ink film and reflected from the surface on which it is printed. Transparent pigments do not hide the background, but rather allow the background material to be seen through the film, e.g. alumina hydrate, magnesium carbonate, calcium carbonate, barium sulphate, clays (in the decreasing order of transparency).

Transparent white pigments are also used to reduce colour strength of inks, to aid dispersion of some of the colour pigments, to help to carry some of the heavier pigments, and to make tints. However, the most important functions of a transparent pigment is to extend the ink and to decrease the concentration of the more costly materials in the formula. So, the transparent pigments are also called "extender pigments".

The chemistry of some important white pigments is discussed in the following paragraphs.

Titanium dioxide (TiO$_2$)   It exists in three crystalline forms. Only anatase and rutile forms have good pigmentary properties. Rutile form is more thermally stable. Anatase transforms rapidly to rutile at a temperature above 700°C. The rutile form is bluer, more opaque and harder than anatase. Titanium dioxide is insoluble, chemically inert, and has excellent resistance to heat and light. Compared to other white pigments, titanium dioxide has the highest refractive index. Rutile titanium dioxide is the most opaque white pigment by virtue of its high refractive index (2.72). Rutile is used to make opaque tints from transparent inks. Because of the less abrasive nature, anatase is preferred in gravure inks.

Calcium carbonate (CaCO$_3$) It is available as ground lime stone or as precipitated calcium carbonate. The precipitated form (known as precipitated chalk) is usually used in printing inks. The precipitated chalk is prepared by passing carbon dioxide into a suspension of calcium carbonate in water. This produces calcium bicarbonate which is unstable and decomposes rapidly when heated, back to calcium carbonate which precipitates out in finely divided form.

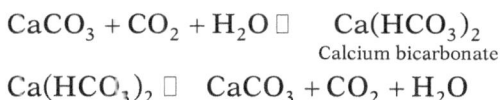

$$CaCO_3 + CO_2 + H_2O \ \square \quad Ca(HCO_3)_2$$
Calcium bicarbonate
$$Ca(HCO_3)_2 \ \square \quad CaCO_3 + CO_2 + H_2O$$

The precipitated calcium carbonate is separated and washed. It is widely used as a general purpose transparent extender.

Clay It is a complex aluminosilicate found abundantly in nature. The hexagonal plate-like form of clay, called Kaolin, is mainly used in printing inks. It is a transparent white pigment and is used largely as an extender for letterpress and screen-printing inks.

Coloured Inorganic Pigments

A large number of simple inorganic salts are coloured but relatively few of them have sufficient colour strength or are stable enough to be useful as pigments. Among the few that are suitable, the most important ones are iron blues.

Iron blues These are the main coloured inorganic pigments used in printing inks. The iron blues are chemically same and are made up of ferric ferrocyanide, $Fe_4[Fe(CN)_6]_3$.

The production of iron blues involves the following steps.

In the first step, a light blue precipitate of ferrous ferrocyanide is produced by the reaction of potassium ferrocyanide with ferrous sulphate solution.

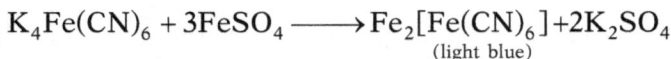

$$K_4Fe(CN)_6 + 3FeSO_4 \longrightarrow \underset{\text{(light blue)}}{Fe_2[Fe(CN)_6]} + 2K_2SO_4$$

The light blue precipitate obtained is then filtered and washed.

In the next step, the light blue precipitate is converted into a deep blue precipitate of ferric ferrocyanide, $Fe_4[Fe(CN)_6]_3$. This conversion is brought about by the addition of an oxidizing agent (such as sodium dichromate or sodium chlorate) and hydrochloric acid. The different shades of iron blues are obtained by varying the conditions of manufacture, i.e., changing the acidity of the solution, the temperature, and the time of heating during the oxidation.

Iron blues have good resistance to light, heat, oils, and solvents, but most have poor alkali resistance. They may be mixed with most other pigments, except those which are alkaline in nature and are reducing agents. The important iron blues are Bronze blue, Milori blue, Prussian blue, and Chinese blue.

Ultramarine blue finds limited applications in printing inks, due to its poor colour strength and texture. The uses of chrome and lead pigments in printing inks have been completely eliminated because of their toxic nature.

Generally inorganic pigments consist of hard dense crystals which in some cases are difficult to grind into a vehicle and which can give inks with poor working properties. Perhaps the greatest advantage of inorganic pigments is their low cost.

REVIEW QUESTIONS

1. What is a dye? What are the basic requirements of a true dye?
2. Discuss the classification of dyes based on their applications.
3. Explain the following terms.
    i.   Chromophores
    ii. Auxochromes
    iii. Chromogens
    iv. Bathochromic groups
    v.   Hypsochromic groups
4. Explain briefly the classification of dyes based on their structure.
5. Nitrobenzene is pale yellow while *p*-nitrophenol is deep yellow in colour. Explain why.
6. Discuss the theory of colour and chemical constitution of dyes.
7. Write a note on nomenclature of dyes.
8. Give the preparation of the following dyes.
    i.   Bismarck brown
    ii. Malachite green
9. How is alizarin manufactured?
10. What are azo dyes? How are they generally prepared?
11. Write a note on:
    i.   Mordant dyes
    ii. Vat dyes
    iii. Reactive dyes
12. Explain the importance of diazotization and coupling reactions in the preparation of azo dyes.
13. Describe the synthesis of the following.
    i.   *p*-Amino azo benzene
    ii. Methyl orange
    iii. Pararosaniline
    iv. Rosaniline

14. What are pigments? Name the types of pigments.
15. Describe the following properties of pigments.
    i. Strength
    ii. Brightness
    iii. Fastness
    iv. Dispersibility
    v. Hiding power and transparency
16. Write short notes on:
    i. Carbon black
    ii. White inorganic pigments
    iii. Flushed pigments
17. Discuss the special features of organic pigments.
18. How are the following pigments prepared?
    i. Pigment red 57:1
    ii. Pigment yellow 13
19. Give the structure of copper phthalocyanine pigment (pigment blue 15).
20. Discuss the chemistry of the following pigments.
    i. Titanium dioxide
    ii. Calcium carbonate
21. What are the advantages and disadvantages of flushed pigments?
22. Describe the preparation of iron blues. How are the different shades of iron blue obtained?
23. Write the structural formulae of the following.
    i. Aniline yellow
    ii. Naphthol yellow S
    iii. Bismarck brown
    iv. Auramine G
    v. Malachite green
    vi. Rosaniline
    vii. Alizarin

ADDITIONAL READING

Finar, I.L. (2004). *Organic Chemistry*, 6th edn. Vol. I. Pearson Education Ltd., Singapore.

Herbst, W. and Hunger, K. (1993). *Industrial Organic Pigments*. VCH, Weinheim, Germany.

Morgans, W.M. (1977). *Pigments for Paints and Inks*. SITA Technology, London.

Patton, T.C. (ed.). (1979). *Pigments Hand Book*. Vol. I, II & III. John Wiley and Sons, New York.

7

VARNISHES AND LACQUERS

VARNISHES

Varnish is traditionally a combination of a drying oil, a resin and a thinner or solvent. In general, varnish is a homogeneous dispersion or solution of natural or synthetic resin in oils and/ or thinner.

Varnishes are commonly used as a protective coating or as a decorative coating. It is an important part of an ink (called vehicle) in which other materials like pigment, drier, wax and modifier are dispersed or dissolved. On drying it forms a transparent, hard, glossy, lustrous and protective film. After being applied, the film-forming substances in varnishes either harden directly, as soon as the solvent has fully evaporated, or harden after evaporation of the solvent through certain curing processes, primarily chemical reaction between oils and oxygen from the air (autoxidation) and chemical reactions between components of the varnish.

Resin varnishes dry by evaporation of the solvent and harden almost immediately upon drying. Acrylic and waterborne varnishes dry upon evaporation of the water but experience an extended curing period. Oil, polyurethane, and epoxy varnishes remain liquid even after evaporation of the solvent but quickly begin to cure, undergoing successive stages from liquid or syrupy, to gummy, to tacky, to "dry to the touch", to hard. Environmental factors such as heat and humidity play a very large role in the drying and curing times of varnishes. In the case of classic varnish, the cure rate depends on the type of oil used and, to some extent, on the ratio of oil to resin. The drying and curing time of all the varnishes may be speeded up by exposure to an energy source such as sunlight or heat. Other than acrylic and waterborne types, all varnishes are highly flammable in their liquid state due to the presence of flammable solvents and oils.

TYPES OF VARNISHES

Varnishes are broadly divided into two main types.

Spirit Varnishes

These varnishes consist of a resin dissolved in a volatile solvent. They are dried quite rapidly by the evaporation of the solvent. But the film formed is more brittle and hence less durable. In addition to this, the film is easily affected by weathering. The typical example is a solution of shellac in alcohol, commonly known as spirit varnish. These types of varnishes are extensively used for polishing wooden furniture.

Oil Varnishes

These varnishes consist of one or more resins (natural or synthetic) dispersed/dissolved in a drying oil and a volatile solvent medium. The oil reduces the natural brittleness of the pure resin film. Oil varnishes are dried or hardened by evaporation of the volatile solvent, followed by oxidation and polymerization of the drying oil. So it takes longer time for drying. However, the film formed is hard, less brittle, quite lustrous and durable. A typical example is copal varnish, which is prepared by dissolving copal in linseed oil and mixing with required amount of turpentine. They are used both in exterior and interior decorative applications.

CONSTITUENTS OF VARNISHES

In addition to the main constituents like resins, drying oils and solvents, varnishes also contain other additives such as driers and antiskinning agents.

Resins They provide hardness, elasticity, durability, chemical resistance, adhesion, weather resistance and water-repellency to the film formed. Resins commonly used in varnishes are natural resins (like shellac, rosin, copal, dammar, amber, etc.) and synthetic resins (like urea–formaldehyde, alkyds, phenol–formaldehyde, urethane, etc.).

Drying oils A drying oil is one that contains a large percentage of fat molecules of unsaturated fatty acids. They dry by oxidation and polymerization. The greater the degree of unsaturation, the faster is the rate of drying. They help in the drying of varnishes. The commonly used drying oils are linseed oil, soybean oil, china wood oil (tung oil) and fish oil.

Solvents or thinners They provide the medium required for the dispersion or dissolution of other constituents (mainly resin) of varnish, and also reduce the viscosity of the medium. The volatile solvent evaporates readily and thus helps in the hardening of varnish. Traditionally, natural (organic) turpentine was used as the solvent or thinner, but has been replaced by several mineral-based turpentine substitutes. Some of the common solvents used are white spirit, naphtha, kerosene, acetone, tolyl, butyl and ethyl alcohols.

Driers It acts as a catalyst to speed up the oxidative polymerization of drying oil constituent of the varnish. For example, dying oil varnishes are usually dried by a slow, complicated chemical reaction called oxidative polymerization. A drier is added to such varnishes to accelerate their drying. It is not changed chemically during the reaction. The most effective driers are linoleates, resinates, tungstates and naphthenates of Co, Mn, Pb and Zn.

Antiskinning agents They are added to prevent skinning of a varnish during its storage and hence they enhance the shelf life of the varnish. Methyl ethyl ketone, tertiary amyl alcohol and hydroquinone are used for this purpose.

SYNTHETIC VARNISHES

Drying-Oil-Modified Varnishes

There are many types of drying oils, including linseed oil, tall oil, soybean oil, and tung oil. These contain high levels of unsaturated fatty acids. By definition, drying oils are not true varnishes though often in modern terms the accomplish the same thing. Drying oils may be modified into varnishes by boiling them for a required time. The temperature and time of cooking determines the viscosity of the varnish obtained. They cure through a reaction between the polyunsaturated portion of the oil and oxygen from the air (oxidative polymerization). Their curing rate can be improved by the use of additives such as driers. However, certain non-toxic by-products of the curing process are emitted from the oil film even after it is dry to the touch and over a considerable period of time. Drying-oil-modified varnishes are used in lithographic, letterpress and screen-printing inks.

Alkyd Varnishes

The modern commercially produced varnishes employ some form of alkyd for providing a protective film. Alkyds or chemically modified vegetable oils, which perform well in a wide range of conditions and can be engineered to speed up the cure rate and thus harden faster. One such varnish is made by the reaction of phthalic anhydride with glycerol to produce alkyd or glyptal resin. The resin so obtained is then reacted with a drying oil (such as linseed oil) to produce a linseed alkyd. The use of a drying oil with long fatty acid chains results in a long-oil alkyd varnish. These varnishes produce durable and flexible film, which is resistant to abrasion and discoloration from UV light, with good adhesion qualities.

Better exterior varnishes employ alkyds made from high-performance oils and contain UV-absorbers; this improves gloss-retention and extends the lifetime of the finish.

Alkyd varnishes are mainly used to make inks for metal decorating purposes. They are also used in lithographic, letterpress and screen-printing inks.

Urethane Varnishes

Polyurethane resins are obtained by the condensation reaction of a diisocyanate with an alcohol. The resin materials on reaction with a drying oil produce a urethane-modified drying oil varnish. These varnishes are also known as "oil-modified" polyurethanes. They typically provide hard, durable, abrasion-resistant and water-resistant film on drying. Polyurethane coatings cure after evaporation of the solvent by a variety of chemical reactions between components within the original mix, or by reaction with moisture from the air. They dry and cure faster than alkyd varnishes.

Compared to simple oil or shellac varnishes, polyurethane varnishes provide harder, tougher and better waterproof film. However, urethane varnishes are light- and UV-sensitive and therefore will turn yellow rather quickly. Hence, they are not suitable for exterior applications. Oil-modified polyurethanes are currently the most widely used wood floor finishes. They are used in sheet-fed lithographic inks.

Phenolic Varnishes

These are made by the reaction of phenolic resins with drying oils. Phenolic resins are the condensation products of alkyl phenols and formaldehyde. During the preparation of phenolic resins, some amount of rosin or rosin ester [obtained by the reaction of rosin with pentaerythritol (2,2-Bis(hydroxy methyl) 1,3-propane diol), $C(CH_2OH)_4$] is added. The addition of rosin improves the oil solubility of the resin and makes the varnish suitable for use in the formulation of inks. Rosin-modified phenolic varnishes are widely used in high-gloss inks.

APPLICATIONS OF VARNISHES

1. Varnishes are widely used in wood-finishing works. They improve the appearance and intensify the ornamental grains of wood surfaces.

2. The synthetic varnishes are primarily being used as the dispersion medium in variety of printing inks.

3. Varnishes are usually applied on metal article as protective coatings against corrosion.

4. Picture varnish is used as a final coating for a painting to protect the picture (from atmospheric oxygen and mechanical abrasion) and to unify the appearance of the surface.

5. Retouch varnish is used to give a full and wet look to the surface of an unfinished painting before work is resumed.

6. Mixing varnish is used as an additive to the painting medium to accelerate the drying time, to add gloss, and to give body to a glaze.

LACQUERS

The dispersion of cellulose derivatives, resins, and plasticizers in solvents and diluents is generally known as lacquers.

Such dispersion dries upon by evaporation of its volatile constituents, and provides protective as well as decorative coating. On drying, they give a tough, durable and glossy coating. They also provide moisture-, heat- and rub-resistant film. Lacquers may be regarded as a modified form of spirit varnish that gives a transparent film. The lacquer added with a pigment is called pigmented lacquer or lacquer enamel.

CONSTITUENTS OF LACQUERS

Lacquers are generally compounded from the following materials.

Cellulose derivatives They provide the required hardness and durability to the film. The commonly used cellulose derivatives are cellulose acetate, cellulose nitrate, and ethyl cellulose.

Resins The resin (natural and synthetic) materials are added to give the required thickness, high gloss, and good adhesion to the film. The principal resins used are copal, phenol–formaldehyde, and alkyd resin.

Solvents They provide a suitable medium for the dispersion of cellulose derivatives and resins, and also help in the drying of lacquers. Solvents commonly employed are acetone, ethyl acetate, methyl ethyl ketone, amyl acetate, dioxane, etc.

Plasticizers The purpose of adding plasticizers is to obtain a smooth and flexible film. They act as internal lubricants to the resin used in the formulation. The most commonly used plasticizers are castor oil, dibutyl phosphate, tricresyl phosphate, and tributyl phthalate.

Diluents (or extenders) They are low-cost liquids which reduce the viscosity of the lacquer medium by acting as thinners. As a result, the lacquer can be easily handled and applied. Important diluents used are petroleum naphtha and tolyl alcohol.

APPLICATIONS OF LACQUERS IN PRINTING TECHNOLOGY

1. A wear-resistant lacquer is added to the image areas during the development of additive pre-sensitized plates.
2. Lacquers are used in some screen-printing inks.
3. Used as coating materials for papers, book and magazine covers, labels, annual reports, and all sorts of product brochures and packaging materials. Nitrocellulose lacquers are commonly used and they provide good gloss and protection.

REVIEW QUESTIONS

1. What is a varnish? What are the different constituents of a varnish and explain their functions?
2. Distinguish between oil varnish and spirit varnish.
3. What are the characteristic features of a good varnish?
4. Write short notes on:
   i. Drying oil varnishes
   ii. Alkyd varnishes
   iii. Urethane varnishes
5. Discuss the applications of varnishes.
6. What is a lacquer? Name the ingredients of a lacquer and discuss their functions.
7. Give the applications of lacquers.

ADDITIONAL READING

NIIR Board of Consultants and Engineers. (2006). *Paints, Pigments, Varnishes and Enamel Technology Hand Book.* Asia Pacific Business Press, New Delhi.

NIIR Board. (2007). *Modern Technology of Paints, Varnishes and Lacquer,* 2nd edn. National Institute of Industrial Research, New Delhi.

NPCS Board of Consultants and Engineers. (2007). *Selected Formulary Book on Inks, Paints, Lacquers, Varnishes and Enamel.* NIIR Project Consultancy Services, New Delhi.

Webb, M. (2000). *Lacquers: Technology and Conservation.* Butterworth-Heinemann, Oxford.

8   PRINTING INKS

INTRODUCTION

Printing inks are involved in almost every aspect of human activity. We are educated, informed, and entertained by books, newspapers and magazines. The money we spend, the stamps we use to send letters, the advertising literature that helps us to make useful buying decisions, all rely on the printing ink for utility and impact. There are four independent components in the printing system:

1. the press, which is the mechanical means,
2. the type (or plate), which provides the message,
3. the paper (or other substrate), which is the vehicle to carry the message and
4. the printing ink, a vital link which ties all of the others together and makes the message visible. The purpose of an ink is to impart information by producing an image on a substrate.

Printing ink is a mixture of colouring matter dispersed in a vehicle (or carrier), which forms a fluid or paste that can be printed on a substrate and dried. The colourants used are generally pigments, toners, dyes or combinations of these materials, which are selected to provide the desired colour contrast with the substrate on which the ink is printed. Printing inks are applied in thin films on many substrates such as paper, metal foils, plastic films, textiles and glasses. Printing inks can be designed to have decorative, protective or communication functions. In some cases, combinations of these functions are achieved.

TYPES OF PRINTING INKS

Generally printing inks vary considerably in their physical appearance, composition, method of application, and drying mechanism. However, most of the printing inks may be classified into five basic types according to the process of printing (such as letterpress, lithographic, flexographic, gravure and screen process) by which they are applied. Based on this, there are five types of inks like, letterpress, lithographic, flexographic, gravure and screen-process inks. In addition to these, a number of specialized inks are also used in different printing processes. For example, some of the speciality types are metallic inks, cold-set inks, watercolour inks and magnetic inks.

However, all these printing inks can be conveniently divided into two main groups, called paste and liquid inks. Lithographic and letterpress inks are called paste inks. They are the high-viscosity inks because their viscosity is much higher. Gravure and flexographic inks are called liquid inks. They are the low-viscosity inks. The paste inks range from the simple news inks containing a large proportion of a non-drying mineral oil, to the litho- and letterpress inks based on a drying oil vehicle. On the other hand, liquid inks contain a large proportion of a volatile solvent.

Screen-printing inks lie somewhere between the two (paste and liquid inks) and have been described as "short and buttery", which is a description of the rheology required by their mode of use.

INGREDIENTS IN PRINTING INKS

Printing inks are mixtures of three basic ingredients such as vehicle, pigment and additives. Miscellaneous other ingredients are also added to impart special properties.

Vehicle

The vehicle consists of a varnish, solvent, and other additives required for giving a proper performance. It is also defined as the mixture of all printing ink ingredients except the pigment. It acts as a carrier for the pigment during the printing process, and also serves to bind the pigment to the substrate. The nature of the vehicle determines the tack and flow characteristics of a finished ink. The selection of vehicle depends on

i. the type of printing process (litho, letterpress, flexo or gravure) involved,

ii. the nature of the substrate (coated paper, newsprint or metal),

iii. the method of drying (heat set, non-heat set, radiation, or air drying), and

iv. other end-use requirements.

Composition of various vehicles

Drying oil vehicle   The main drying oils used in lithographic, letterpress and screen-printing inks are linseed oil, soybean oil, rosin oil and china wood oil (tung oil). However, the raw drying oils are not suitable as printing ink vehicles and hence they must be modified. The drying oils are usually modified by boiling. As a result of boiling, the viscosity and other properties of drying oil are radically altered. The temperature and time of cooking determine the viscosity of the varnish. Ink made up of drying oil vehicle usually dry by the oxidation of drying oil content. The mechanism of oxidation is rather complicated, which involves the absorption of oxygen by the drying oil, and the presence of a drier (as catalysts) speeds up this action.

The modified linseed oil varnishes have excellent melting properties for most pigments. They have good transfer and drying qualities and provide good binding on paper. Varnishes produced by the combination of drying oils and synthetic resins are used to obtain faster and harder drying inks.

Non-drying oil vehicle   This vehicle consists of non-drying, penetrating oils (such as rosin oils, petroleum oils, etc.), which are modified with various resins to impart suitable tack and flow characteristics. Inks (such as news and comic inks) containing this type of vehicle usually dry by the absorption of the vehicle into the paper.

Solvent–resin vehicle   Inks with solvent–resin vehicles (such as gravure and flexographic inks) dry by the evaporation of the solvent content. In gravure inks, the vehicles are usually made up of a combination of gums and modified rosins with low boiling hydrocarbons as solvents. Flexographic inks may contain either alcohols or other volatile solvents, in combination with suitable resins or gums. Generally, the nature of the substrate to be printed determines the choice of specific ingredients.

Heat-set inks for letterpress and offset are formulated from varnishes containing rosin esters and hydrocarbon resins dissolved in high boiling petroleum solvents. The selection of the proper synthetic resins for heat-set inks is very important, so as to provide proper solvent release, good gloss and rapid drying.

Inks for aluminium foil, boxboard, plastic films, etc. are formulated with varnishes containing lacquer type solvents and resins.

Resin–oil vehicle   It consists of a balanced combination of resin, oil and solvent. The solvent is rapidly absorbed by the paper, leaving a partially dry ink film of resin and oil which subsequently hardens by oxidation. Quick-set ink vehicle often contains rosin-modified phenolics, alkyds, maleic-rosin esters or hydrocarbon resins. They usually contain high boiling aliphatic hydrocarbon solvents (heat-set solvents). When the quick-set ink is printed on a coated paper, the solvent soaks into the coating, leaving the high viscocity vehicle on the surface. The viscocity rises rapidly and the ink sets by oxidative drying.

Glycol vehicle   These are made from resins that are insoluble in water, but soluble in glycol type solvents. Moisture-set or precipitation inks with glycol vehicle dry by precipitation of its binder (resin) rather than by evaporation of its solvents. The addition of water in the form of steam or water spray to the ink system causes the precipitation of the resin, along with the pigment.

"Fast drying" glycol type inks are prepared using a vehicle composed of an acidic resin, neutralized by an amine, and dissolved in a suitable glycol. During printing, the resin salt decomposes and the amine is almost completely absorbed by the substrate. The resin, which in its acidic form is insoluble in the glycol, precipitates and forms a hard film within a few minutes without the use of moisture.

Resin–wax vehicle   The inks made with resin–wax vehicles dry by cold setting and contain a high percentage of waxes and resins. "Cold-set" or "hot-melt" inks are solid at room temperature and are liquefied by heating. These inks can be used only in fountains on presses specially designed for their use. Rollers, plates and fountains are heated to keep the ink in the liquid form. The ink sets on contact with the cold surface of the paper. Such inks have limited commercial applications.

Water-soluble gum vehicle   The vehicle consists of a water-soluble gum (such as gum arabic) and other similar materials dissolved in water and glycerine or other water-miscible solvents. Watercolour inks are made from this type of vehicle and are primarily used for novel applications.

Photoreactive vehicle   The vehicles for UV-curing inks contain an unsaturated prepolymer (oligomer), polyfunctional monomers and monofunctional monomers which cross-link on exposure to UV light. A photoinitiator is required to start the cross-linking process. On absorption of energy from

UV light, the photoinitiator (such as benzoin methyl ether) forms free radicals thus initiating the polymerization and cross-linking. The prepolymer oligomers used may be derived from by combining acrylic acid with urethane, epoxide or polyester resins. The resins provide the tough and glossy properties, whereas the unsaturated nature of the acrylate portion helps in the cross-linking of the prepolymer resins. Monomers (as reactive diluents) with low viscosity are used to reduce the high viscosity of the prepolymer resins and to assist in pigment wetting to produce workable ink vehicle. Additives such as reactive or non-reactive plasticizers are added to give flexibility to the cured film. Amine synergists (such as tertiary amine triethane amine, $(CH_3CH_2)_3N$) are added to accelerate the reaction. The UV inks are essentially non-volatile, and cure very rapidly on exposure to UV radiation. Thus curing takes place without the use of heat and with practically no emission of volatile matters (since they do not contain solvents). UV-curing inks are particularly used on lithographic printing on non-absorbent substrates such as plastics and tin plates. As a recent innovation, UV-drying water-based inks are also been used for screen-printing applications.

The vehicles for EB inks are also formulated with the same type of oligomers used in UV inks, but without the photoinitiator. The energy of electro-beam is sufficiently high to produce free radicals, on collision with the molecules in binder, and hence no photoinitiator is required.

Solvents in ink   Solvents are liquids that are capable of dissolving other substances. Generally the solvents used in printing inks are able to dissolve resins or oils in the ink vehicle. Diluents are liquids that may not completely dissolve the resin by itself. Solvents can also be thinners, but most often thinners are blends of solvents and diluents. Reducer is another name for the thinner, referring to the solvent blends used to reduce the viscosity of the ink to running viscosity.

Table 8.1 Some common solvents used in printing inks

Chemical name	Comparative drying time	Boiling range (°C)	Flash point* (°C)	Uses
Hydrocarbons				
Hexane	7	66–70	–18	Flexo and gravure inks
Toluene	26	109–112	7	Flexo and gravure inks
Xylene	88	135–143	27	Flexo and gravure inks
Alcohols				
Ethyl alcohol	18	75–80	18	Flexo and gravure inks
Isopropyl alcohol	27	81–83	21	Flexo and screen-printing inks
Esters				
Ethyl acetate	10	72–80	6	Flexo, gravure and screen-printing inks
Isopropyl acetate	12	84–90	16	Gravure and screen-printing inks
Glycol ethers				
Cellosolve	190	132–137	54	Flexo, gravure and screen-printing inks
Ketones				
Acetone	5	56–57	–9	Flexo and gravure inks
Methyl ethyl ketone	11	78–81	–1	Flexo, gravure and screen-printing inks

* The flash point is the lowest temperature at which a solvent will form a vapour/air mixture, which will produce a flash, i.e., explode weakly, without burning continuously.

The selection of a suitable solvent (Table 8.1) for a particular ink mainly depends on its solvent power, that is, its effectiveness in dissolving a particular material such as oil, resin, or wax. The rate of evaporation (or drying time) of the solvent also becomes important, since the solvent is usually required to leave the ink film immediately after printing. Usually a mixture of solvents is used to maintain the correct rate of evaporation. Other properties of solvent which have to be considered are their flashpoint, toxicity, odour, and their tendency to leave a trace of solvent trapped in the dry film.

Pigments

The colouring materials used in printing inks are generally pigments rather than dyes. Pigments are finely divided coloured substances which are dispersed in the vehicle. They (organic or inorganic) provide the desired colour in inks. They are also responsible for some of the specific properties of the inks, such as specific gravity, opacity or transparency, viscosity, yield value, and resistance to light, heat and chemicals. Pigments should have the combination of properties that are required in a printing ink. The most important of these properties are good colour strength, reasonable stability to light and chemicals, fine particle size and the ability to be dispersed in normal ink vehicle to give ink with good flow properties. Judicious selection of pigment is also required, according to the use of the ink, in considering subsequent operations such as varnishing, waxing, lacquering, or laminating.

Toners are undiluted full strength pigments that are used to strengthen weak batches of pigments. Occasionally, dyes are used as toners.

Requirements of ink pigments   Pigments used in printing inks should have the following characteristics.

Good colour strength   This is essentially required when thin films of ink are applied, in offset and flexographic printing.

The important characteristics controlling the colour of a printing ink are the colour strength of the pigment and its opacity.

Good light fastness The pigments used in the ink formulation should not fade on exposure to sun or UV light. This is essentially required where the printed end product is likely to be exposed to UV light for a longer time (such as posters, packaging products, etc.).

Stability towards chemical attack This is the main requirement of the packaging industry where pigments may come in contact with materials which may be acidic (such as foodstuff containing citrus acids, vinegar, etc.), or alkaline (such as soaps, detergents, etc.).

Fine particle size This provides the required optical and colour properties to the final ink film. The colour strength of a pigment increases with decreasing particle size.

Dispersibility in normal ink vehicle The surface energy or wettability of the pigment should be compatible with the vehicle used in ink. For successful wetting of the pigment particles, the surface energy of the pigment must be higher than that of the dispersant (surface tension).

Pigments used in printing inks The most common pigments used in printing inks are as follows:

Black pigments Carbon black is used in inks for newsprinting, publication, commercial and packaging printing.

White pigments Opaque white pigments commonly used in inks, in the order of decreasing opacity, are titanium dioxide, zinc sulphide, lithopone, and zinc oxide.

Transparent white pigments (extenders) commonly used in printing inks, in the order of decreasing transparency, are aluminium hydrate, magnesium carbonate, calcium carbonate, talc and clay. Extenders are sometimes used to reduce the colour strength and to change the rheology of inks.

Inorganic colour pigments   Iron blues (Milori blue and Prussian blue) are used as toners in some black inks.

Organic pigments   Some important types of organic pigments used widely in printing inks are the following.

Yellow pigments   These pigments are used in all types of printing inks, e.g. diarylide yellow and Hansa yellow.

Orange pigment   It is a transparent pigment and not very fast to light, e.g. diarylide orange.

Red pigments   The para reds and toluidines are fairly fast to light and are semi-transparent. They are used in poster and label inks. Lithol Rubine reds are widely used as magentas in all types of four colour printing.

Blue pigments   Phthalocyanine blues have excellent light fastness and excellent resistance to most chemicals. They are used as cyans in all four colour printing processes.

Victoria blue is fairly fast to light, but generally have greater colour strength and are less expensive. They are useful for toning black ink and are used in printing packaging.

Purple pigments   Methyl violet is widely used for toning black inks. Carbazole and Vat violets are very expensive, but are used in packaging where performance and resistance properties are required.

Green pigments   Malachite green is used in inks for printing packaging materials. Phthalocyanine green is useful as a green toner.

Fluorescent pigments   They are particularly used where brilliant fluorescent shades are required. These types of pigments are mainly used in silk screen, gravure, flexographic, letterpress and offset inks. They are mainly used in printing packaging materials, greeting cards, wrapping paper and posters.

The important classical pigments that are extensively used in multicolour printing include cyan (copper phthalocyanine blue or Pigment blue 15), magenta (calcium 4B toner or Pigment Red 57:1) and yellow (benzidine yellow or Pigment Yellow 13).

Additives

Other ingredients such as driers, waxes, antioxidants, lubricants, gums, plasticizers and surfactants are added to impart certain special characteristics to inks.

Driers Driers are used in inks that dry by oxidation. They catalyse the oxidation and subsequent drying of the varnish. Hence, they reduce the drying time of inks. Driers are generally soaps of cobalt, manganese, and other metals formed with organic acids such as linoleic acid and naphthenic acid. These are used in two different forms, as liquid driers or as paste driers. Paste driers are salts of fatty acids, while liquid driers are salts of aliphatic acids obtained by oxidizing petroleum.

Its normal range of concentration in inks is between 0.5% and 4%. Higher concentration of driers causes the ink to skin and dry on the press, to fill in half-tones and in many cases slow down drying of the prints causing the sheets to stick. In lithographic inks, excess of drier may lead to greasing and scumming. When more water is added to reduce it, the ink and paper become waterlogged. This results in slower drying. The proper selection of the drier and the amount of drier to be used are more important in the formulation of oxidizing printing inks. The pigments also influence the drying properties of the inks.

Waxes Waxes are added to printing inks to improve slip, mar resistance, and water repellency. They usually decrease gloss and hence must be used with care. Addition of wax will shorten an ink. Polyethylene, microcrystalline, and paraffin wax are commonly used in printing inks. The wax may be cooked directly into the varnish, or prepared as a compound and added directly to the ink. Wax compounds may also be added to letterpress and lithoinks to reduce their tack without much affecting their flow properties.

Antioxidants (Antiskinning agents)   They are added to some inks to prevent the skinning of ink in the fountain and on the press rollers. Antioxidants retard the premature oxidation of inks on the press rollers when used at low concentrations. Excess amount of antioxidants may prevent the ink from drying on the paper after printing. A good antioxidant is one that increases the skinning time greatly without much increase in drying time. Some of the widely used antioxidants are methyl ethyl ketoxime, butylated hydroxy toluene and hydroquinone.

Lubricants   Lubricants are added to reduce the tack of an ink and helps it to set quickly. They also lubricate the ink for the proper distribution and transfer of ink. Excessive lubrication may cause an ink to become greasy and leads to poor printing.

Gums   They increase the viscosity of inks. They 'pull' the ink together and help it to print sharply. In lithographic inks they help to overcome emulsification, improve drying and prevent chalking of the ink.

Plasticizers   These are used in most ink systems that dry by solvent evaporation such as gravure and flexography. The commonly used plasticizers are vegetable oils and dibutyl phthalate. They are usually added to impart flexibility to ink film. Such inks are particularly used in flexible packaging applications.

Surfactants   They are used as wetting and dispersing agents for pigments in printing-ink vehicles.

PROPERTIES OF INKS

The main properties of inks are drying, rheology (flow property), colour, and end-use properties.

Drying

Drying may be defined as any process that results in the transformation of a fluid-printing-ink into a solid film. An ink is considered dry when a print does not stick or transfer to another surface pressed into contact with it. The proper drying of an ink is very important.

Depending on the type of ink and printing process, the drying of printing inks is accomplished by one of the physical or chemical methods such as absorption, solvent evaporation, oxidation and polymerization, precipitation, cold setting and radiation curing.

Absorption   Certain inks (e.g. oil-based newspaper inks) dry by penetration or absorption into the pores of the printed stock. This is accomplished by the gross penetration of the ink vehicle into the pore of the substrate, by the partial separation of the vehicle from the pigment, and by the diffusion of the vehicle throughout the paper. The ability of an ink to penetrate into paper depends on the number and size of the pores present in the paper, the affinity or receptivity of the stock for the ink, and the mobility of the ink. Absorption drying can be successful only if there is a proper balance between the oil absorbency of the paper and the viscosity of the ink. Insufficient absorption can cause slow drying and leads to ink set off to the underside of the next sheet. This can be avoided by applying a thinner film of ink. If too much of the vehicle is absorbed into the paper, unprotected pigment is left on the surface and this may easily rub off or set off on to the following sheet. This effect is more common in coated papers. The excessive absorption of a vehicle can also lead to show through, making the printed image visible from the other side of the sheet. The main advantage of absorption drying is that it takes place rapidly and the ink dried by this method maintains the highest press speed.

Unfortunately a print produced with an ink which has largely been absorbed into paper is bound to have a dull and poor appearance.

Solvent evaporation   Drying of ink can also take place by the evaporation of a volatile solvent content in the ink vehicle. A printing ink which dries solely by solvent evaporation requires a vehicle composed of synthetic resins dissolved or dispersed in suitable high boiling hydrocarbon solvents. After printing, the solvent evaporates leaving the resin binding the pigment to the surface of the paper. The removal of solvent can be easily achieved by application of heat and/or air blow. Web offset and letterpress inks which dry by solvent evaporation are usually called heat-set inks. Flexographic and rotogravure inks usually contain more volatile solvents than heat-set inks and may not require much heat to dry.

Evaporation is the most important method in the drying of gravure and flexographic inks, particularly when they are applied to non-absorbent substrates like aluminium foil and plastic films. The choice of suitable solvent for such inks mainly depends on the vapour pressure of the solvent, because the rate of evaporation of a solvent depends on its vapour pressure at the temperature of drying. Other factors which may influence the choice of solvents are their odour (in the case of inks for packaging), the nature of the printing plates (which may be attacked by some solvents), and the nature of material being printed. In the case of plastic film, it is necessary to include a solvent which will slightly dissolve the surface of the film to achieve an adequate bonding of the ink.

When slow drying occurs in inks that dry by solvent evaporation, the drying should be adjusted by adding a more volatile solvent to the ink or by increasing the drying oven temperature.

Oxidation and polymerization   Drying of inks containing drying-oil varnishes usually takes place by oxidation and polymerization of drying oil molecules. The oxygen present in air adds on to the double bonds of the molecules of the drying-oil to form hydroperoxides. The so-formed hydroperoxide decomposes in the presence of a catalyst (called drier) to form free radicals. These free radicals bring about the cross-linking of drying-oil molecules. The oxygen is continuously absorbed and the viscosity increases until a hard, dry film is obtained. During the process of drying, certain simple molecules such as water, carbon dioxide, acetic acid and formic acid are given out as vapour. In this method, both the concentration of oxygen and that of the drier become equally important.

Precipitation Drying of an ink can also be caused by the precipitation of its binder. This can be accomplished by adding a diluent, such as water in the form of steam, to a hygroscopic solvent ink system. This causes the solubility of the resin in the film to decrease sharply and causes it to precipitate when its tolerance for the diluent is reached. Further drying is accomplished by absorption of the solvent into the stock and then by evaporation.

Moisture-set inks dry by this method. The solvents used in the vehicles are glycols. The resins generally employed are maleic or fumaric acid-modified rosin.

Cold setting (or hot melt) In this method, inks are applied to a substrate in a molten state and upon cooling they form a dry ink film. Cold-set inks dry by this method. They consist of pigments dispersed in plasticized waxes having melting points ranging from 65°C to 95°C. They are used on presses with fountains, rollers and plates which are heated above the melting point of the inks. The inks are melted and maintained in a fluid condition until they are impressed on the relatively cold paper where they revert almost instantly to their normal solid state.

Radiation curing In this method, drying of ink is brought about by the use of different forms of radiant energy such as ultraviolet, electron beam, infrared and microwave energy.

UV inks are composed of pigments, oligomers, reactive (acrylated) resins, monomers, photoinitiators and other additives. By the absorption of UV energy, photoinitiators undergo haemolytic cleavage to form free radicals. These free radicals initiate the polymerization reaction and results in curing of ink. The curing of ink can also takes place by cationic polymerization reaction. This technology uses vinyl ethers and epoxy resins for the oligomers, reactive resins, and monomers. The initiators form Lewis acids upon absorption of the UV energy and these acids cause cationic polymerization. The cationic mechanism has improved adhesion and flexibility, and offers lower viscosity compared to free radical mechanism. But the cationic mechanism is very sensitive to humidity conditions and amine contamination.

EB inks are dried by the use of energy from electron beam generator. They are very similar to acrylic UV inks without the photoinitiators. The inks are therefore cheaper and more stable in storage. The energy of the electron beam causes the acrylate double bonds to cleave directly, forming free radicals which initiate polymerization. EB systems require an inert gas (nitrogen) curing chamber in order to prevent the inhibiting effect of oxygen on drying. Due to this, EB applications are usually used for web printing. EB curing units are relatively costly compared to UV units. But they require much less energy to operate, are sensitive to colour and ink film thickness, and produce much less heat. EB inks are cheaper compared to UV inks and are more stable in storage due to the absence of photoinitiators. Unlike UV inks, EB inks are "colour-blind" and hence they do not dry one colour faster than another.

Infrared and microwave inks are respectively cured by the application of IR and microwave radiation energy. The radiant energy causes the inks to heat and dry through partial evaporation of solvent. The drying mechanism also involves the absorption of the ink into the porous substrate. By the use of IR radiation, there is possibility of thermal degradation of paper-coating ingredients due to the absorption of heat in the non-image areas.

Rheology (Flow Property)

The rheology or flow of inks is their primary physical property. In printing industry, the flow properties of inks will have a direct impact on their performance. The properties of inks vary according to the materials used in their formulation. They may be long or short. A long ink is one that can be stretched out into a long string when it is pulled away from the ink slab with an ink knife or a finger. A short ink breaks sooner under the same conditions. Generally long inks have good flow characteristics.

Lithographic inks are required to have good flow in order to transfer well from one roller to another and to tolerate the emulsification of water. Screen inks need to be short so that they will not form strings when the screen is pulled away from the print. An offset ink, however, must not be too long, since this can result in ink flying or misting. Long, stringy inks which tend to cobweb between the rollers are usually prone to flying.

The common terms used to describe ink rheology or flow property are viscosity, yield value, thixotropy, and tack.

Viscosity   It is the property of a liquid or fluid by virtue of which it offers resistance to its own flow. It is expressed in units, Ns m^{-2} (formerly poise, P, where 1 Ns m^{-2} = 10 P). In a Newtonian liquid, any stress produces a flow, and the rate of flow is proportional to the stress. But inks are generally non-Newtonian and have a non-linear flow curve.

Low-viscosity inks are preferred for fine-line flexography and shallow-cell highlights in gravure printing. The viscous nature of litho- and letterpress inks has made press designers to use a multitude roller in the ink distribution unit to ensure uniform thin films and proper transfer of ink to the printing plate. Gravure and flexographic inks are the low-viscosity inks. The ink-distribution systems of flexo- and gravure printing presses are very simple having few inking rollers. In most cases, higher press speeds require low-viscosity inks and vice versa. Printing on smooth, dense solids can be best achieved using high-viscosity inks.

Different types of viscometers are available for the precise measurement of viscosity of different types of inks. These include capillary, falling body, rotational and other types. Ink rheology measurements are made using modern instruments such as the rotational torsion, cone-plate, falling rod, and efflux viscometers.

Yield value   The yield value of an ink is the force per unit area required to initiate it to flow. When a pigment is added to a Newtonian liquid, the solid particles increase the resistance to flow and the viscosity increases. Most litho- and letterpress inks contain a large amount of pigment. Such inks will not flow at all until the force applied is greater than a certain minimum, which is called the yield value. If the yield value of an ink is too high, it will not flow out of the ink fountain.

Thixotropy   Thixotropy is exhibited when the viscosity of a substance decreases with time on applying a shearing stress at a constant rate. The viscosity then increases when the shearing action is removed. A thixotropic ink is one which has a false body, but can be easily returned to a more fluid state with agitation. Many letterpress and lithographic inks have the property of building up resistance to flow, which is broken down by shearing or agitation, but which rebuilds itself when ink is allowed to stand. This effect is known as thixotropy. The ink manufacturer should be asked to control this effect as much as possible.

Tack   The ability of an ink to flow is obviously essential for its effective performance as it moves from the duct via the rollers and the printing surface to the paper. In this process, a film of ink has to split each time when it passes from one roller to the next, ink roller to the plate, and finally in the transfer to the paper.

The tack of an ink may be defined as the resistance of splitting offered by an ink during the different stages of the printing process. The tack of an ink can be measured roughly by a simple finger test in which the ink is dabbled out on a slab. The value of this simple test is considerably increased if the tack of the ink is compared with that of a sample ink of known tack. Unfortunately, the tack of a normal printing ink is liable to change during long periods of storage, and to overcome this, specially formulated tack pastes (e.g. Mander-Kidd) are available. Tack of an ink is commonly measured using an inkometer.

A high-tack ink results in heat generation in the rollers, high power consumption, and picking or tearing of the paper or the pulling of the coating from coated paper. However, a high-tack ink generally prints sharper by letterpress and cleaner with lithography, where the water uptake of the ink becomes important. A low-tack ink may cause poor dot sharpness, drying problems, and low gloss.

To obtain proper trapping of colour in wet multicolour printing, the first down colour usually should have a higher tack than the second down and so forth. On lithographic presses, the printing plates are kept moist with water. Some of the water becomes emulsified in the ink, thus decreasing its tack.

In the case of solvent-containing inks, the tack continuously increases due to the following factors.

- the evaporation of some solvent from the ink;
- the absorption of some solvent by the rollers or blankets;
- the heating effect that takes place during press operation which leads to further evaporation of solvent.

Colour

The key property of all types of ink is its colour. Pigments provide the characteristic colour in printing inks. Colour can be defined as one of the virtual experiences or one of the impressions we receive through our eyes. Colour has three different attributes described as hue (shade), saturation (chroma), and lightness (value).

Hue is that quality of colour by which one expresses the strength, or the amount of a given hue in a colour, or the intensity of the hue. For example, the difference between a greenish yellow or a tan, and a strong brilliant yellow.

Lightness is that quality which expresses the similarity to a series of grays ranging from black to white, such as the difference between a strong red and a light pink, or the difference between any strong or dark colour and a pastel shade of the same colour.

End-Use Properties

The end-use properties are those considerations that determine how printed substrates function through all processing and usage from the time of printing throughout the useful life of the printed products. The ink film formed on the substrate after drying should possess certain essential properties for the effective working performance of the printed product. Some of these required properties are rub-resistance, heat and light resistance, and resistance to materials such as water, solvents, acid, alkali, soap, detergent, oils and fats and waxes. The ink manufacturers should know the end-use requirements of the printed product for any particular job.

Rub-resistance The printed product employed in packaging applications should possess high rub-resistance. Rub-resistance is the measure of the ability of a printed product to withstand abrasion during its working conditions. The printed packaging materials usually experience high degree of abrasion when they may be rubbed against metal parts on package making or package filling machines and subsequently rubbed against the surface of other print in the transportation of filled packages in outer containers. Rub-resistance may be observed by the press operator by rubbing two prints at equal pressure against each other. This is an indication of how well the ink will slip. However, this type of test is not satisfactory because it is impossible to standardize the pressure and the speed of the rub. Standard methods using suitable rubcmeters can be employed to measure the rub-resistance of a printed product.

Some of the possible causes for poor rub-resistance are the following.

- ink is too soft
- stock is very absorbent
- not enough binder in the ink for the particular stock used
- the vehicle portion of the ink penetrates into the stock too much, leaving only the pigment on the surface

The overprint varnishes can be used to protect the ink film from scratching and abrasion. Using an ink with a harder resin and additional wax can improve rub-resistance.

Heat and light resistance   The printed product used for food packaging purposes should have reasonable resistance to heat. The beer or food containers usually require sterilization or pasteurization. So, inks used for printing them should not change colour or deteriorate when subjected to high temperatures.

Inks used for printing posters, showcards and other display materials require good light fastness (resistance to light). Although no print can be completely permanent, the ability of different inks to withstand prolonged exposure to light varies considerably. This is due to the different light fastness of the pigments used in the inks. However, it has been found that the light resistance of a pigment does depend on the vehicle in which it is dispersed and the pigment concentration. The light resistance of printed products can be determined by a standard method using a xenon lamp as the light source.

Resistance to water, solvents and other materials Depending on the nature of working environment, the printed products may come in contact with various materials. So, it is essential that the printed products should be resistant to the action of such materials. For example, print on a soap wrapper should be resistant to materials like water, alkali, and soap. Laminating inks should withstand the solvent in the adhesive.

INK AND WATER INTERACTION

Water does not dissolve in ink, but it does disperse or emulsify in tiny droplets into the ink. This produces a water-in-ink emulsion and the phenomenon is known as emulsification of water in inks. The tendency of the ink to emulsify water is determined by all of the raw materials in the ink, but predominantly those in the vehicle. If an ink emulsifies too much water or emulsifies water in large droplets, it becomes short and buttery and fails to transfer well. This condition can lead to a washed-out print and piling of the ink on the rollers and plate. This can also cause gum streaks when a plate containing such a "water-logged" ink is gummed up.

Some water always becomes emulsified in a lithographic ink when it is being run on a press. Otherwise, the ink will not transfer properly on the ink rollers and a stripping condition will result. The water breaks up into tiny droplets that are dispersed throughout the ink, producing a water-in-ink emulsion (emulsified water). Water droplets can also deposit on the surface of inked image areas of the plate (surface water).

Excessive surface water in ink can lead to slow drying, snow flaky prints, piling of ink on the rollers, pigment flocculation, and poor trapping in multicolour wet printing. If sheets run on a lithographic press are covered with a more or less uniform tint of the ink, the press operator is faced with a tinting problem. One cause of tinting is the formation of an ink-in-water emulsion. If an ink-in-water emulsion is obtained, the dampening solution is said to be dirty, and it carries ink droplets on the plate's non-image areas. These are transferred to the blanket and then to the paper where they tint the press sheets.

The tack of ink increases when water is completely emulsified in an ink. On the other hand, surface water greatly reduces the apparent tack of the ink.

REVIEW QUESTIONS

1. What are the main ingredients of a printing ink?
2. Explain the functions of vehicles and pigments in printing inks.
3. Discuss the functions of the following additives in printing ink.
   - i. Driers
   - ii. Waxes
   - iii. Antioxidants
   - iv. Lubricants
   - v. Gum
   - vi. Surfactants
4. Discuss the types of printing inks.
5. Discuss the special features of the following types of varnishes.
   - i. Solvent–resin varnish
   - ii. Resin–oil varnish
   - iii. Non-drying oil varnish
   - iv. Photo-reactive varnish
   - v. Resin–wax varnish
6. What are the basic requirements of a pigment to be used in printing inks? Explain.
7. Discuss briefly the types of pigments used in ink manufacture.
8. Write a short note on solvents used in printing inks.

9. What is meant by drying of an ink? Mention the different mechanisms involved in the drying of inks.

10. Describe the mechanism of drying of ink by absorption method. What are the merits and demerits of this method?

11. Explain the following mechanisms of drying of ink.
    i. By oxidation and polymerization
    ii. By precipitation
    iii. By cold setting
    iv. By radiation curing

12. Discuss the mechanism of drying of ink by solvent evaporation. Give the importance of this mechanism.

13. Explain the following flow properties of ink.
    i. Viscosity
    ii. Yield value
    iii. Thixotropy
    iv. Tack of ink

14. "The tack of solvent-containing inks continuously increases during the printing process". Explain why.

15. What do you mean by the following terms?
    i. Long and short inks
    ii. Yield value of inks
    iii. Tack of inks

16. What are the adverse effects of a high-tack ink?

17. Write a short note on end-use properties of ink.

18. Discuss the significance of the following end-use properties of ink.
    i. Rub-resistance
    ii. Heat and light resistance

19. Write an explanatory note on ink and water interaction.

20. Explain the adverse effects of the following.
    i. Emulsification of water into ink
    ii. Emulsification of ink into water

ADDITIONAL READING

Bisset, D.E. *et al.* (eds.). (1988). *The Printing Ink Manual*. Chapman and Hall, London.

Eldred, N.R. and Scarlett, T. (1990). *What the Printer Should Know About Ink*, 2nd edn. Graphic Arts Technical Foundation, USA.

Fetsko, J. (ed.). (1983). *Raw Material Data Hand Book*. Vol. 1–4. National Printing Ink Manufacturers Association, Hasbrouck Heights, N.J.

Leach, R. (ed.). (1993). *Printing Ink Manual*, 5th edn. Blueprint.

NAPIM (1987). *Test Methods for Printing Inks*. National Association of Printing Ink Manufacturer, Harrison, New York.

NAPIM (1988). *Printing Ink Hand Book*. National Association of Printing Ink Manufacturer, Harrison, New York.

9 PAPER TECHNOLOGY

INTRODUCTION

Paper was inverted by Tsai Lua, a Chinese, in AD 105. In the beginning, paper was used only for handwriting. Later on after the invention of printing, paper was also used in printing. The word **paper** comes from the Greek term for the ancient Egyptian writing material called **papyrus**, which was formed from beaten strips of papyrus plants. Papyrus was produced as early as 3000 BC in Egypt, and sold to ancient Greece and Rome. Today, paper includes a wide range of products with varied applications in different fields such as communication, cultural, educational, artistic, hygienic, sanitary, as well as for storage and transportation of all kinds of goods. It is almost impossible to imagine a life without paper.

Paper is a commodity of thin material produced by the amalgamation of fibres, typically vegetable fibres composed of cellulose, which are subsequently held together by hydrogen bonding. While the fibres used are usually natural in origin, a wide variety of synthetic fibres, such as polypropylene and polyethylene, may be incorporated into paper as a way of imparting desirable physical properties. The most common source of these kinds of fibres is wood pulp from pulpwood trees, largely softwoods and hardwoods, such as spruce and aspen respectively. Other vegetable fibre materials including those of cotton, hemp, linen, and rice may also be used.

It is hardly difficult to imagine a printing industry without paper. Although there is considerable amount of printing being carried out on plastic and metal substrates, paper is likely to be the printers' most important basic material for many years to come. It is clearly essential for anyone involved in printing to have a basic knowledge of the chemistry of paper and technology involved in the manufacture of paper.

THE NATURE OF PAPER

The paper consists essentially of a mat or web of intermeshed cellulose fibres held together by hydrogen bonding. It is the presence of hydrogen bonding in cellulose that provides the fibres with the ability to bond together into a strong sheet.

PAPER FIBRES AND PULPS

Cellulose fibres can be regarded as the common building bricks of plant architecture. A tree trunk is made up of solid bundles of fibrovascular tissues formed in continuous concentric rings. Cellulose fibres suitable for paper-making are usually extracted from the fibrovascular tissue of the plant, in which the cell walls making up the fibre are cellulose. The fibres are hollow tubes glued together with lignin. The materials called hemicelluloses coexist with lignin. These materials are contained in the region (called middle lamella) between the adjacent fibres and they bind together the fibrils. The surface of a fibre wall is composed of long threads called fibrils. These fibrils can be split into even finer threads called microfibrils, which are made up of linear chains of several hundred cellulose molecules arranged in parallel bundles. The fibre wall consists of four distinct layers as shown in Figure 9.1. The primary outer layer is very thin (~ 50 nm) and consists of randomly arranged fibrils. In the secondary layers (S_1, S_2 and S_3), the fibrils are arranged in parallel helical patterns. The inner secondary layer (S_2) is the thickest (200–800 nm) and contains about 80–95% of the cellulose fibre.

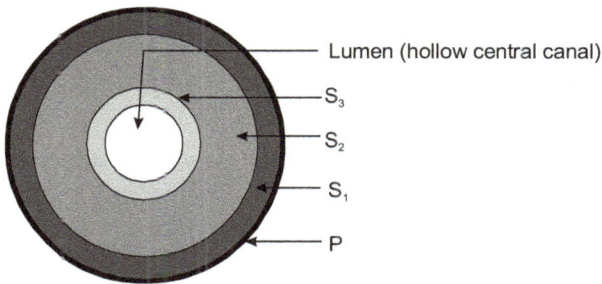

P—Primary layer (~50 nm); S$_1$—Outer secondary layer (70–200 nm); S$_2$—Inner secondary layer (200–800 nm); S$_3$—Tertiary secondary layer (70–80 nm)

Figure 9.1 Structure of a wood fibre

Cellulose is a polysaccharide having the general formula $(C_6H_{10}O_5)_n$. It is a crystalline linear polymer with the repeating units of glucose held together by 1, 4-linkages.

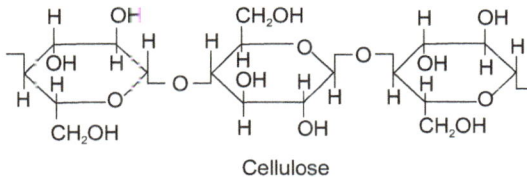

Cellulose

The polymer molecules vary in length depending upon the type of fibre. Thus, cotton fibres may have up to 5000 repeat units in length, whereas wood pulps have maximum up to 2000 units. The presence of hydroxyl groups in their structure is responsible for the hydrogen-bonding nature of cellulose fibres, which gives them the ability to bond into strong-matted structures. This also gives rise to the hygroscopic nature of cellulose. The vegetable fibres usually contain other materials also along with cellulose. In a few fibrous materials, like cotton and linen, the cellulose exists in a pure form but in most plants it is mixed with appreciable quantities of other materials like lignin (a polymeric, phenolic compound that serves to bind the cellulose fibres together). Lignin is stiff and brittle; it turns yellow with age and on exposure to light. From the paper-makers' point of view, these non-cellulose materials of plants are undesirable. Hence, to produce a high-quality white paper, it is necessary to remove the lignin during the pulp preparation step.

Today, about 80% of the world's paper is made from wood pulps. Cellulose fibres are obtained from different plant sources, which include coniferous softwoods such as pine, spruce, larch and redwoods; non-coniferous hardwoods and softwoods such as beech, poplar, eucalyptus and birch; bast fibres such as linen, hemp and jute; seed hairs such as cotton; grass fibres such as straw, bagasse and bamboo and leaf fibres such as esparto, sisal and manila.

The fibres extracted from these plant sources differ in length, breadth and shape but are similar in their general structure. Coniferous softwoods such as pine and spruce provide longer fibres with good strength characteristics. However, these long fibres may produce paper sheets with unsmooth surfaces. Deciduous woods such as birch and eucalyptus produce shorter fibres which are lower in strength but produce smoother papers. Thus the choice of fibres influences the nature and quality of the paper produced. Certain properties of the pulp also depend upon the process employed to separate the fibres from the source material.

PAPER MANUFACTURE

The paper manufacturing process involves the following main stages.

1. The preparation of cellulose pulp from wood or other plant origin.

2. The preparation of stock for paper-making machine, which involves the process of beating or refining and the inclusion of additives.

3. The paper-making operation in which water is progressively removed from the pulp first by drainage

through the wire mesh and then by suction, pressure and the application of heat to form a continuous web of dry paper.

4. The paper-finishing processes which may follow paper-making include calendering, coating and glazing. These treatment methods are applied to improve or modify the surface characteristics of the paper.

PULP PREPARATION

The main objective of the pulp-making is to separate the fibres from the raw material source so that they may be reformed into a sheet of paper. This can be achieved by exploiting the differences in both physical and chemical properties of cellulose and lignin. Fibres can be separated chemically, mechanically, or via a combination of the two.

The mechanical method works by breaking the boundary between the fibre and the lignin containing middle lamella. The chemical method brings about the separation of fibres through dissolution of lignin that binds the fibres together.

The main types of pulp-making processes are

1. Mechanical (groundwood) process
2. Chemical process
3. Combined chemical and mechanical processes

Mechanical (Groundwood) Process

In the mechanical (groundwood) pulping process, debarked logs are fed into grinders where they are pressed against rotating grinding stones in the presence of water, and are fibreized to form a slurry containing minute particles of both fibrous and non-fibrous portions of the wood. The frictional heat generated by the process softens the lignin, allowing the fibres to separate. However, mechanical pulping does not remove the lignin from the fibrous material. Mechanical pulp is usually produced from spruce wood.

The disadvantages of the mechanical pulping process are as follows:

- The grinding process causes extensive damage to the fibres and hence produces a paper with poor strength.
- The damage to the fibres creates a great deal of fines that leads to slow drainage of the sheet on the wire mesh during the paper-making process.
- The lignin and other impurities in the pulp result in a paper with a poor colour, that discolours further on exposure to sunlight, and it becomes brittle with age.

Despite these disadvantages, mechanical pulps have an important role in paper-making. Mechanical pulps are much cheaper than those produced by chemical methods, since the pulp yield is very high (about 95%) and no expensive chemicals are used. It is suitable for newspapers, paper towels and most magazines, which have a very short working life, and in which strength, good colour and permanence are not essential properties. The lignin that is present in mechanical pulp contributes high opacity to the paper, so that light-weight publication papers often contain mechanical pulp. To counteract their poor strength and colour, mechanical pulps are normally blended with a chemical pulp.

A large variety of mechanical pulps (generally called "alphabet" pulps) are now being produced by special mechanical separation methods that are assisted by heat or chemicals. Some of these mechanical pulps are produced by refiner mechanical pulp (RMP) and thermo-mechanical pulp (TMP) processes. These pulps usually produce a strong paper without the addition of chemical pulps.

Refiner mechanical pulp (RMP) is produced by passing wood chips or even wood shavings and sawdust between the grooved discs of a refiner. The disc bars compress the wood particles generating heat, which softens the lignin and shears the fibre apart. The resulting RMP contains a large proportion of long fibres with less damaged fibres than ordinary stone-ground wood pulp. So, it leads to the formation of a strong paper. The energy consumption of the RMP process is higher than that required for the stone ground wood pulp process. The added advantage of the method is that it is possible to produce ground wood pulp from sawdust or wood shavings instead of wood logs.

In the thermo-mechanical pulp (TMP) process, wood is chipped and then fed into large steam-heated refiners where the chips are squeezed and fibreized between two steel discs. Refining takes place in the presence of steam, and when combined with fractional heating, it softens the lignin and allows fibre separation to occur. The mechanical wood pulps so obtained are subsequently passed through a suitable screen to remove the unseparated fibre clumps. The fibres produced in this way are more completely separated from each other and suffer less damage than those obtained by stone-grinding process. The process gives a higher yield of fibre than the chemical pulping process.

Thermo-mechanical pulps are stronger and cleaner than ground wood pulps. Many useful papers such as newsprint can be produced from TMP without blending with a chemical pulp.

Chemical Process

The purpose of a chemical pulping process is to break down the chemical structure of lignin and to render it soluble in the cooking liquor so that it may be washed away from the cellulose fibres. Because lignin holds the plant cells together, chemical pulping sets free the fibres and makes pulp. Thus the lignin can be washed away leaving long undamaged fibres in a pulp that is free of contaminants. The chemical pulps usually produce papers that are whiter, brighter, stronger and more stable to light and ageing than that produced by mechanical pulps. In the process lignin can be dissolved by using a variety of materials. Generally, solutions of sulphites, sulphides and alkalis are used because they are cheap.

The wood chips are placed in a pressure reactor and cooked with a suitable material so as to dissolve lignin. The materials used also dissolve most of the hemicelluloses that coexist with lignin (in the middle lamella). Hence, a low yield of pulp (about 50%) is obtained. On digestion, the wood chips are converted into a soft pulp that consists mainly of cellulose fibres. After the completion of the process, the pressure is released from the digester and the pulp is washed. The undigested knots are screened out and the pulp is finally bleached. In most chemical processes small amounts of anthraquinone are usually added (0.05–0.2%) so as to improve the rate of delignification and the yield.

Chemical pulps tend to cost more than mechanical pulps, largely due to the low yield, 40–50% of the original wood. Since the process preserves fibre length, chemical pulps tend to make strong papers. Another advantage of chemical pulping is that burning the lignin removed during pulping produces the majority of the heat and electricity needed to run the process.

The following are the chemical processes involved in the pulp preparation.

Kraft process　Most chemical pulp is made using the Kraft process. The word **Kraft**, which means "strength" in German reflects a principal characteristic of this pulp (i.e., a long-fibred strong pulp). It was developed by Carl Dahl in 1884 and is now used for about 80% of production volume of paper.

In the kraft process (also known as kraft pulping or sulphate process), the raw materials used are sodium sulphate (Na_2SO_4) and calcium carbonate ($CaCO_3$) mixed with water. The wood chips are added with a required amount of this mixture and then heated by passing steam under pressure in a large pressure vessel called digester. The process is commonly carried out in a continuous digester (such as the kamyr digester). Under the prevailing conditions, sodium sulphate is reduced to sodium sulphide (Na_2S) which in turn is hydrolysed to sodium hydroxide (NaOH) and the lignin is digested.

As the sodium hydroxide is used up in the digestion process, it is replenished by the hydrolysis of sodium sulphide. This controlled supply of sodium hydroxide is advantageous as it avoids drastic alkaline conditions that would attack cellulose and yet allows the digestion to continue at a steady rate. In this way, the fibre length is retained and a strong pulp is formed. The carbonate reacts with the resins in wood, forming water-soluble soaps. Hence, the major amounts of non-fibrous materials are dissolved and separated from the cellulose fibres.

The kraft process is particularly useful for resinous woods like pine that cannot be extracted effectively by the sulphite process. The spent extracted pulping liquor, called black liquor, is concentrated by evaporation and burnt in the recovery boiler to recover the inorganic chemicals for reuse in the pulping process. The recovery boiler also generates high pressure steam for the mill processes. The inorganic portion of the black liquor is used to regenerate the sodium hydroxide and sodium sulphide needed for pulping in a process called causticizing.

In the case of softwood (coniferous) pulping, soap-like substance is collected from the liquor during evaporation. The soap is acidified to produce tall oil, a source of resin acids, fatty acids and other chemicals. Also turpentine originates from softwood.

The high-pressure steam, from the recovery boiler, is led to turbogenerators, reducing the steam pressure for the mill use and generating electricity. A modern kraft pulp mill is more than self-sufficient in its electrical generation and normally will provide a net flow of energy to the local electrical grid. Additionally, bark and wood residues are often fired in a separate power boiler to generate steam.

Kraft process differs from the sulphite process by using alkaline solution, which is less corrosive to the equipment. Sulphite process cannot process pulp from all wood species, for example from pine. Kraft process is also more efficient than the sulphite process. It produces stronger fibre, however the fibre is also rougher and darker that makes it somewhat more challenging to bleach.

After kraft pulping, the pulp can be used directly for bags and boxes or further delignified, during bleaching, to produce white pulp for printing and writing.

Sulphite process The raw materials used in the sulphite process are iron sulphide and calcium carbonate (limestone). These are used to produce the acid–sulphite liquor, i.e., a mixture of calcium bisulphite and free sulphur dioxide in water, which is required for the digestion. The acid–sulphite liquor is prepared by spraying water down a tower packed with limestone and by blowing sulphur dioxide gas up the tower.

In sulphite process, the woodchips are treated with a solution of calcium bisulphite and free sulphur dioxide. The mixture is digested in a digester (a large cylindrical pressure cooker with an acid-resistant lining) by passing steam at a pressure of about 700 kPa for about 20 hours. During digestion process, much of the lignin and other non-fibrous materials pass into the acid–sulphite liquor. The sulphurous acid produced from liberated sulphur dioxide dissolves the lignin.

Wood materials commonly used for making sulphite pulp are spruce, hemlock and balsam. Since most of the resins in the wood are not dissolved during pulping, this process is ineffective on resinous woods such as pine. The sulphurous acid also attacks the outer wall of the fibre resulting in a weak paper. The recovery of the raw materials from spent acid–sulphite liquor is difficult. The spent acid–sulphite liquor also causes severe waste disposal and stream pollution problems. Since the outer wall of the fibre is softened, the pulp wets up more easily. This is advantageous during paper-making process. However, it produces brighter pulps because of the removal of larger amount of lignin.

Combined Chemical and Mechanical Processes

In recent years, a number of pulping methods have been developed, which involves the techniques of both chemical and mechanical processes. They usually involve a mild chemical digestion followed by mechanical disintegration. As one might expect, the pulp produced by these combined processes is intermediate in quality in between a mechanical and chemical pulp. They produce generally high-yield pulp with inferior ageing properties compared to purely chemical pulps and superior to that of purely mechanical pulps. These pulps may be suitable for making packaging products, or may require further bleaching to produce printing papers.

The **semi-chemical** process is a two-stage process in which the normal chemical pulping is followed by mechanical action. The neutral sulphite semi-chemical (NSSC) process is used mainly for mixed hardwoods. The digestion of the wood chips is carried out with sodium sulphite and sodium carbonate (or sodium hydroxide), which is followed by refining.

In **mechanical–chemical** methods, the pulp is prepared by agitating the wood materials in presence of a suitable chemical. For example, wood materials such as straw or bagasse are agitated at high speed in a hot dilute solution of sodium hydroxide.

In chemi-thermo-mechanical pulp-making (CTMP) process, the wood chips are softened by treating with sodium sulphite before they are defibred at a high temperature. The pulp produced by this process is much stronger than the thermo-mechanical pulp (TMP).

Screening and Cleaning

The cooked pulp is then washed and screened to achieve a more uniform quality. Most of the dissolved impurities are removed by washing the pulp with water. The washed pulp usually contains impurities such as knots, pieces of bark, uncooked fibre bundles, sand, grit, and resin that would weaken the paper. Hence, these impurities present in the pulp must be removed before paper-making.

Screening is a multi-stage operation in which diluted pulp is made to pass through a series of perforated plates or screens (with holes of different sizes). The separated fibres easily pass through the holes, while the large-sized materials such as knots, bark pieces and fibre bundles are retained by the screens.

Cleaning follows the screening process, in which grit and other materials are separated by differences in density. This can be performed by allowing the diluted pulp to flow over sand traps, into which heavy particles are settled and retained. However, more efficient cleaning is possible by using centrifugal and vortex cleaners. In the vortex cleaners, the diluted pulp is forced under pressure into an inverted cone. The rotating motion applies a centrifugal force to the constituents, the magnitude of which is related to the mass of the particles. Pure pulp spirals down the wall of the cone and then rises upwards in an inner vortex to leave by a central outlet at the top. The heavier, denser non-cellulose particles are carried downwards to the open point of the cone where they are continuously discharged.

Bleaching

Bleaching may be considered as an extension of digestion process and that removes the residual lignin and coloured organic impurities in the fibres. Bleaching of pulp is essential for producing whiter, brighter and light-fast papers. Different types of bleaching procedures and chemicals are commonly used for this purpose.

Most industrial bleaching methods are based on the use of chlorine and its compounds (such as sodium or calcium hypochlorite) as oxidizing bleaching agents. The chlorine reacts with the residual lignin and the chlorinated lignin obtained is then removed by treatment with sodium hydroxide solution. But the residual traces of chlorine left in the pulp may slowly degrade and weaken the cellulose fibre.

The sulphate wood pulps usually contain appreciable amounts of residual lignin which are difficult to remove in one-stage bleaching. Multi-stage bleaching is usually employed in such cases to give high whiteness without any loss of strength. Each stage uses a different form of chlorine as bleaching agent. In the first stage, the unbleached pulp is reacted with chlorine gas, followed by sodium hydroxide extraction. The pulp, in the next stage, is treated with an aqueous solution of calcium hypochlorite and then extracted with sodium hydroxide solution. The final bleaching is carried out with chlorine dioxide (ClO_2) in water, followed by sodium hydroxide extraction.

The brightness of mechanical pulps is increased by bleaching them either with hydrogen peroxide (H_2O_2) or sodium hydrosulphite ($NaHSO_3$). These bleaching agents remove colour without removing the lignin. Hydrogen peroxide bleaching in conjunction with sodium silicate is used on recycled pulps, chemical pulps, TMP and CTMP. The mechanism of bleaching involves the reaction of the perhydroxy anion (HO_2^-) with coloured organic impurities. The controlled decomposition of hydrogen peroxide in alkaline medium gives the required perhydroxy anion.

$$H_2O_2 + NaOH \longrightarrow HO_2^- + H_2O$$

The added sodium silicate maintains the required pH (9–10) in which controlled decomposition of hydrogen peroxide is possible.

In the case of a plant designed to produce pulp to make brown sack paper or linerboard for boxes and packaging, the pulp does not always need to be bleached to a high brightness. In these cases, a high yield of fibre from wood can be achieved.

After the bleaching process, the pulp is then delivered to the stock preparation plant.

STOCK PREPARATION

The processes of stock preparation can be considered in two distinct steps such as

1. Breaking or defibreing
2. Beating or refining

Breaking or Defibreing

This involves redispersion of pulp sheets into a suspension of the individual fibres in water. The process is commonly carried out in a hydrapulper, an open cup-shaped tank with a diameter of about 6 m and an average depth of 3 m. A high-speed propeller is fixed at the bottom of the tank. Both the propeller and the lower walls of the tank are fitted with metal blades. The propeller revolving at a speed causes severe agitating action forcing the pulp sheets and water to move up and then into the swirling centre of the tank. During this movement, the action of metal blades disintegrates the sheets of pulp. A perforated plate is fixed somewhere at the bottom of the tank, which allows the fibres to pass out into a storage chest.

Beating or Refining

This involves certain physical modifications of the fibres that results in the formation of a paper with desired characteristics. In practice the non-fibrous additives of the pulp are also added at this step.

This is the most important process in paper-making, for the reason that it determines the physical characteristics of the final paper. Paper made from an unbeaten pulp is usually weak and flabby because of the poor adhesion between the fibres. Technically, the process can be performed either by bath process of beating or by continuous process of refining. In recent years, most paper mills have adopted the continuous process of refining.

The properties of fibres are greatly changed during the refining process. This is mainly due to the breakage of the intrafibre hydrogen bonding and fibrillation of the secondary walls. This results in the formation of interfibril hydrogen bonding between the fibres. The beating usually breaks down the water-resistant outer wall of the fibres, thereby exposing the inner fibrils. This effect is called fibrillation. The fibres slowly take up water and swell due to continuous fibrillation. As a result, the fibres become more hydrated and stronger. The refining process renders the fibres more flexible by bruising them so that they may deform more easily and be drawn closer together by surface tension forces during drying of the paper web, thus bringing the microfibrils within hydrogen bonding distance. Increasing the refining, further reduces the fibre rigidity. Increased refining results in the production of paper sheets having higher density, bursting and tensile strength, and folder endurance. This is mainly due to the stronger internal bonding of the fibres.

Extensive refining can also cause disadvantages. In addition to the high cost of power to run the refiner, the tear strength and absorbency of the paper are reduced. As fibres are shortened by cutting, the resistance to tear quickly reaches a maximum and then decreases. Absorbency gradually decreases because the shorter fibres give rise to a denser packing.

Pulp Additives

The non-fibrous additives (such as sizing agents, fillers and colouring materials) are continuously added into the stock flow at convenient points before the pulp is introduced to the paper-making machine.

The main non-fibrous materials that are used as additives in the stock preparation process include water, sizing agents, fillers (loadings) and colouring materials.

These additives are incorporated in the stock so as to meet the following main requirements:

- to improve the properties of the fibres in the stock,
- to improve the paper-making process, and
- to produce high-quality papers with desired characteristics.

Water   Water is the main ingredient (about 95%) in the preparation of stock for paper-making. In paper industry, water is usually obtained from different sources like rivers or borewells. Thus the quality of water and its control becomes an important factor. The water used must be clean because coloured water would affect the brightness and whiteness of the paper. The presence of any suspended matter such as silt can cause spotting. It should be free from salts of iron, manganese and other transition metal ions that would cause discoloration or interference with hydrogen bleaching. The water sample should be reasonably soft. Therefore, it is essential to have a water-treatment plant in each paper mill so as to get pure water.

Sizing agent   Sizing agent is a substance that is applied to fibres during paper manufacture in order to curb their tendency to absorb liquids by capillary action. By doing so, sizing keeps the ink on the surface of the paper where it is intended to remain. In addition, sizing affects the other properties of paper such as abrasiveness, finish, printability, smoothness, surface bond strength, surface porosity and fluffing.

There are two major types of sizing.

1. Engine (rosin) sizing where sizing agent is added to the pulp before paper-making and is commonly known as internal sizing.

2. Surface sizing where the surface of paper is treated with sizing agent either on the paper machine or as a separate process after paper-making.

Rosin is the common sizing agent used for internal sizing. It is a hard amber-coloured resin obtained from pine trees and is insoluble in water. Rosin size is obtained by the boiling of rosin with sodium hydroxide solution. Thus rosin is converted into rosin soap, that is soluble in water. The rosin size is added to the pulp mixture and it is then "set" by adding paper-makers' alum which is aluminium sulphate $[Al_2(SO_4)_3]$. The sulphuric acid produced by the hydrolysis of aluminium sulphate neutralizes the alkaline soap and precipitates the rosin on to the fibres.

$$Al_2(SO_4)_3 + 6H_2O \longrightarrow 2Al(OH)_3 + 3H_2SO_4$$

The rosin particles so formed soften and fuse during the drying stage of paper-making. The coating of sizing agents on the fibres makes them more water-resistant, and reduces the penetration of water into the paper by capillary action. Where a high degree of water resistance is required, wax emulsions are used either separately or in conjunction with rosin size.

However, the acid produced by hydrolysis of residual alum may reduce the life of the paper obtained. The use of acid sizing agents also limits the choice of fillers (such as chalk) because the acid produced may decompose them. To overcome these problems, neutral synthetic sizing agents (such as alkenyl succinic anhydride and alkyl ketone dimer) are now being used for paper-making.

The surface sizing applied to the drying paper web performs different functions such as improving finish, reducing fluffing and controlling ink penetration. The widely used surface sizing agents are starch, gelatin, polyvinyl alcohol and carboxymethyl cellulose.

There are three categories of papers with respect to sizing;

1. unsized (waterleaf),
2. weak sized (slack sized), and
3. strong sized (hard sized).

Waterleaf has low water resistance and includes absorbent papers for blotting. Slack-sized paper is somewhat absorbent and includes newsprint, while hard-sized papers have the highest water resistance.

Fillers (Loadings)   These are finely divided materials which are added to the pulp mainly to improve the opacity of the finished paper. An ideal filler must be chemically inert and available in a finely divided form at a reasonable rate. In addition to this, the material should have a high refractive index. Addition of fillers mainly reduces the cost of a paper since they are much cheaper than fibres. The added fillers usually fill the space between the intermeshing fibres. They improve the printing characteristics of a paper by increasing its smoothness, opacity, brightness, and dimensional stability.

Fillers also cause some adverse effects. As the percentage of filler added increases, the tensile, burst and tear strength of the paper decreases. Abrasive fillers can cause wear on paper machine and printing plate. So, they must be used carefully only in required quantities. Most printing papers normally contains 10–20% filler, whereas writing paper may contain only 5–10% of fillers.

Some of the fillers used are china clay (a naturally occurring aluminosilicate), titanium dioxide (TiO_2), and precipitated chalk ($CaCO_3$). The most widely used filler is china clay or kaolin, since it is relatively cheap. Titanium dioxide is a stable, brilliant white pigment with a very high opacity. It is more expensive but its effectiveness as a filler more than justifies its use. Even small proportions of the pigment have an appreciable effect on the brightness and opacity of a paper. However, it is only used in special applications where high opacity coupled with high paper strength is desirable.

The papers with calcium carbonate as fillers have an excellent affinity for ink. Since calcium carbonate is highly absorbent, it accelerates the setting or vehicle absorption of inks, which is important for fast-running printing presses. Calcium carbonate also contributes to paper permanence, since it neutralizes the acid residues in the paper.

Colouring materials The colouring materials are added to the paper stock either to produce coloured papers or to improve the whiteness of paper. Paper having a slight yellow tint due to residual lignin may be whitened by adding small amounts of blue dye or pigment. The two main types of colouring materials used in paper-making are soluble inorganic pigments and water-soluble dyes.

Pigments may be considered as coloured loadings filling the interspaces between fibres. They are the particulate solids having high opacity as well as colour. They provide good light fastness and chemical resistance to the finished paper. But they can weaken the paper sheet, produce two-sideness and are difficult to disperse evenly. In comparison to dyes, pigments have better resistance to light and chemicals. The commonly used mineral pigments are Prussian blue, ultramarine blue and chromes.

The water-soluble dyes are easy to use and can be metered continuously into the stock. However, they are pH sensitive, less light fast and may require a mordant to fix them to the fibre. Mostly synthetic dyes are used in paper-making and they fall into three main classes: acid, basic and direct.

A suitable colouring material is chosen based on the type of fibre, end-use of the paper and the required shade and light fastness. The synthetic direct dyes are commonly used and the pigments are particularly used where highest opacity is required.

Fluorescent dyes, called optical brighteners or optical bleaches are also used as colouring materials in paper-making. They are used to improve both the whiteness and brightness of paper. These materials are able to absorb invisible UV rays and re-emit them as visible light in the blue region of the spectrum.

In order to impart certain special properties and effects, some other additives are also added. Plasticizers (such as glycerine) are added to obtain a soft and less brittle paper. Defoamers (like mineral oils and fatty acids) are able to disperse air bubbles at the wet-end by reducing the surface tension. The air bubbles generally cause low-density spots in the paper. Slimicides (such as organic cyanate compounds) are added to avoid slime formation by bacterial contamination.

THE PAPER-MAKING PROCESS

After cleaning, refining and inclusion of additives, the diluted pulp is now ready for paper-making. In the paper-making machine, water is continuously removed from the diluted pulp so that the fibres come together to form a thin dry paper web.

A paper-making machine consists of three main parts, namely, the wet end, the press section and the drier section. The diluted pulp introduced at one end (wet end) of the machine is released at other end (drier section) as a dry web. During this journey down the machine, water is removed from the pulp by the natural drainage, suction, pressure and finally by the action of heat.

The stock (diluted pulp containing 99% water) is directed from a head box through a slice which evenly spreads the pulp on to a woven wire mesh moving in a continuous belt. Water is drained through the wire, which is assisted by a series of hydrofoils, suction boxes and rollers. During drainage the fibres are drawn together to form the paper web that contains about 80% of water.

The paper web is then fed into the press section, where more residual water is removed by squeezing the web between synthetic felts on large rollers, followed by suction. In this step the water content of the web is reduced to about 60–65%.

The web then enters the drier section, where most of the remaining water is removed by passing the web over steam-heated drums. Thus the water content of the web is finally reduced to about 5–10%.

The dried web is then subjected to surface sizing in a size press, where a thin layer of starch or carboxymethyl cellulose solution is sprayed on to the paper surfaces to improve ink control and fluff resistance.

PAPER-FINISHING PROCESS

The processes which are usually carried on paper in order to improve its surface finish include calendering and coating. The term finishing refers to the application of the required surface finish to the paper.

Calendering

This process involves the pressing of a paper between heavy rollers. The main objectives of calendering are the following.

- to ensure proper final print quality by imparting smoothness and gloss
- to improve sheet uniformity in thickness, particularly in the cross-machine direction of the paper web.

This is achieved by passing the paper web through a series of steam-heated rollers (rolls) that apply compression and thus imparts smoothness to the paper.

When only a low or medium finish is required, then this can be achieved by calendering the paper on-machine. Some paper machines are equipped with calender rolls between the drier and the reeling sections. Papers that have been finished by machine calenders only are called machine-finished or MF papers. But if a high finish is desired then the paper is supercalendered as a separate process after the paper-making process.

As the degree of calendering is increased, smoothness and gloss of paper are also increased. However, the thickness and stiffness of paper are reduced. Calendering lowers opacity considerably and decreases brightness somewhat by compacting fibres closer together and reducing the light scattering property of the paper. Porosity and ink absorbency are decreased by calendering.

The smoothness and thickness of the paper obtained depends upon the speed of travel (dwell time in the nip), the nip pressure, the moisture content of the paper and the temperature of the rollers. As the temperature of calendering is increased, the smoothness for a particular thickness increases. Smoothness increases with an increase in moisture content. The paper (like glassine paper) made from well-beaten fibres containing high loadings usually gives the best calendering results because the fibres are more flexible and bruised and easier to roll out. Blotting paper is not usually calendered, because the process would reduce its bulk and absorbency. The gloss imparted decreases as the speed of calendering increases because the dwell time (i.e., the time the paper spends between the nips) is insufficient for the heat to penetrate the paper and plasticize it.

The two common types of calendering, which are currently in use, are hard and soft calendering. The hard calenders contain polished hollow iron rolls, which can be steam heated internally to give a roll temperature of about 200°C. The temperature gradient calendering between rolls with one heated to 200°C and the other to 40°C gives improved calendering properties. The main purpose of this is to retain the thickness of the paper and hence its stiffness while imparting surface smoothness. The technique is particularly successful with soft-nip calendering. The heat supplied to the paper plasticizes the surface layers, which enables them to flow under pressure into a smooth formation. The nip pressure should be maintained at a constant level to ensure uniformity of calendering. The long heavy rolls usually sag under the action of gravity resulting in variations in sheet thickness across the roll. This roll deflection can be controlled by using special rolls such as crown and swimming rolls. The main disadvantage of hard calendering is that the paper web obtained may not be uniform in density. This affects the rate of ink absorption and setting during printing. This problem can be reduced in soft-nip calendering.

In soft-nip calendering, the paper web is passed between a pair of rolls, with one being hard (iron) and the other covered with a layer of soft, flexible elastomer. Soft calendering gives paper with uniform bulk density and consequently more even ink density. It gives higher gloss for coated papers and greater tensile strength for a given thickness compared to hard-nip calendering because bonds and fibres are weakened by hard-nip treatments.

Supercalenders comprise alternate hard and soft rolls. The soft rolls are made either from cotton or compressed paper. They are highly polished and larger in diameter than the steam-heated iron rolls. In supercalendering process, the paper web is fed through a vertical bank of rolls which are even in number ranging from 8 to 80, depending on the degree of calendering required. The rolls rotate at different speeds, the faster roll being highly polished, applies a polishing action to the surface of the sheet. Thus the rolls impart a high finish to the paper.

Coating

The surface properties of a paper can be greatly improved by the application of a thin coating of fine mineral pigment mixed with a binder. The coated paper usually will have improved smoothness, gloss and ink receptivity, which are the essential requirements to get good quality print.

The coating material is basically a combination of a fine mineral pigment and a binder (adhesive). The finely divided mineral pigments require an adhesive binder to hold them together and to fix them to the paper. The mineral pigments required are of the same types as those used for loadings but are usually of much better quality and finer particle size. The commonly used mineral pigments are specially refined grades of china clay, precipitated chalk (calcium carbonate) and barium sulphate (blanc fixe). Titanium dioxide (TiO_2) is expensive to use as a coating material but it is often blended to improve the whiteness and opacity of cheaper minerals. The mineral pigment is thoroughly dispersed in a suitable binder or adhesive.

The binders used in coating materials include natural materials like starch and casein, and synthetic copolymers like styrene–butadiene, butadiene–acrylonitrile and butadiene–methylmethacrylate. Normally the synthetic copolymers are used in the form of aqueous dispersions (known as lattices) and in combination with casein or starch. Other additional additives that may be used include a defoaming agent, formaldehyde as a hardener and a wax emulsion to improve the finish.

The paper can be coated either on-machine or off-machine. Off-machine coating is faster but requires an additional investment in drying equipment. The coating characteristics are dependent on the surface properties of the paper. It is essential that the base paper should be flat, reasonably smooth and free from surface faults. More important is that the paper must be uniform in composition and texture; otherwise the coating and any subsequent printing will show similar irregularities. The method of coating also influences the relationship between the smoothness of the coating and that of the base paper.

For instance, in air knife process, the coating is of fairly uniform thickness and so tends to follow the contours of the base paper.

The main methods of application of coating include air knife coating, blade coating and cast coating.

On-machine coaters have a blade or air knife coater installed about one-half to two-thirds of the way down the drier section of the paper machine. An air knife coater applies an excess of coating to the paper web and removes excess coating with a controlled jet of air. A blade coater removes the excess coating with a doctor blade. Blade coating fills the valleys in the paper surface and results in a very smooth coated surface.

A smooth finish is usually obtained when the coating process is followed by supercalendering. However, the cast coating produces the smoothest and glossiest finish. The principle of cast coating is to bring the wet coating applied to one side of a paper in close contact with the highly polished surface of heated chromium drum. In this way the mirror like finish of the chromium-plated drum is transferred to the coated surface of the paper as it dries in contact.

TYPES OF PAPERS

Papers may be classified in different ways based on their basis weight, colour, end use, raw material used, surface treatment, etc. Papers can be graded in 'n' numbers of ways and if we count all permutation and combination of grades, total grades may well exceed 10,000. Some of the major grades of classifications are given in Table 9.1.

Table 9.1 Classification of paper based on various parameters

Parameter	Type	Characteristics
Based on basis weight		
	Tissue	Low weight, $< 40\,g/m^2$
	Paper	Medium weight, $40–120\ g/m^2$
	Paperboard	Medium high weight, $120–200\ g/m^2$
	Board	High weight, $> 200\ g/m^2$
Based on colour		
	Brown paper	Unbleached paper
	White paper	Bleached paper
	Coloured paper	Bleached and dyed or pigmented paper
Based on usage		
	Industrial	Packaging, wrapping, filtering, electrical, etc.
	Cultural	Writing, printing, newspaper, currency, etc.
	Food	Food wrapping, candy wrapping, coffee filter, tea bag, etc.

Based on raw material		
	Wood	Contain fibres from wood.
	Agricultural residue	Fibres from straw, grass or other annual plants.
	Recycled	Recycled or secondary fibre.
Based on surface treatment		
	Coated paper	A paper which has a special coating applied to its surface.
		Materials such as clay, casein, talc, etc. are applied by means of roller or brush applicators; or plastics applied by means of roll or extrusion coaters.
	Uncoated paper	A paper with no coating applied to its surface.
	Laminated paper	A paper built up to a desired thickness or a given desired surface by joining together two or more webs or sheets. The paper thus joined may be alike or different; a totally different material, such as foil, may be laminated with paper.

(Contd.)

Table 9.1 (Continued)

Parameter	Type	Characteristics
Based on finish		
	Fine/Coarse	
	Fine paper	Uncoated writing and printing grade paper including offset, bond, duplicating and photocopying.
	Coarse paper	Various grades of papers used for industrial applications rather than cultural purposes.
	Calendered/ supercalendered	
	Calendered paper	The paper which is calendered on the paper machine.
	Supercalendered paper	The paper which is supercalendered as separate process after the paper-making process.
	Machine finished/Machine glazed	
	Machine finished (MF)	Smooth paper calendered on the paper machine.
	Machine glazed (MG)	Paper with a glossy finish on one side produced on the paper machine by a Yankee cylinder.
	Glazed (Glossed) paper	Paper with high gloss or polish applied to the surface either during the process of manufacture or after the paper is produced, by various methods such as friction glazing, calendering, plating or drying on a Yankee drier.

PROPERTIES OF PAPER

Paper may be characterized by properties like physical properties, strength properties, optical properties, chemical properties and printing properties.

PHYSICAL PROPERTIES

The physical properties of a paper are grammage (basis weight), caliper (thickness), bulk, wire side and felt side, machine- and cross-direction, curl, formation, friction, dimensional stability, compressibility, smoothness, porosity and absorbency.

Grammage (Basis Weight)

The basis weight or grammage is the mass per unit area of a sheet of paper. It is expressed as the weight in grams per square metre (g/m^2). Traditionally, it is defined as the weight of paper per ream. A ream is a specified area of paper, which is frequently expressed in terms of a specified number of sheets (usually 500) of a certain dimension.

Typical grammage values (accepted trade tolerance $\pm 5\%$)

Paper grade	g/m^2
Cigarette tissue	22–25
Newsprint	40–50
Bond	60–90
Paperboard	120–300

Paper is sold by weight but the buyer is interested in the area of paper. The basis weight is what determines, how much area the buyer gets for a given weight. For example, if basis weight is 50 g/m^2, for every 1 kg weight, the buyer gets 20 m^2. When the basis weight is expressed as ream weight, it tells the buyers how many reams he/she gets for a given weight.

Grammage is an important contributor to sheet thickness, stiffness and opacity and determines converting yield. In the printing of web papers, control of basis weight is most important. The total printable area of roll paper is frequently estimated from the gross weight of a given number of rolls and the basis weight of paper. The basis weight is measured either directly during paper manufacturing with the β-ray gauge or off-machine by weighing a precisely cut piece of paper with a balance (TAPPI Standard T 410). The Technical Association of the Pulp and Paper Industry (TAPPI) has developed standards for most techniques that are used to test the properties of paper. These standards establish the specifications and procedures that are necessary before data from testing labs around the country can be comparable. Electronic balances are always preferred to maintain speed, accuracy and precision of measurement.

Caliper (Thickness)

The normal thickness of paper is known as its caliper. It is expressed in micrometers (μm). The caliper of paper is measured by using a micrometer with circular contact surfaces of 16 mm (0.63 inch) diameter as described in TAPPI Standard T 411. Caliper is a critical measurement of uniformity. Variations in caliper can affect several basic properties including strength, optical and roll quality. Thickness is important in filing cards, printing papers, condenser papers, saturating papers, etc. It is usually related directly to the moisture content, the pressure applied at the wet end and the extent of calendering. Monitoring caliper is a useful aid for controlling the nip pressures during calendering. Variations in thickness of paper during a print run will lead to differences in impression and hence in ink density and dot gain.

Typical thickness values (accepted trade tolerance ±10%)

Paper grade	μm
Newsprint	60–80
Label paper (79 g/m^2)	63
Tracing paper (90 g/m^2)	78
Office/business paper	105–110
Tissue (28 g/m^2)	125
Blotting paper (230 g/m^2)	540–590

Bulk

It is another very important parameter of paper particularly for printers. Bulk is a term used to indicate volume or thickness in relation to weight. It is the reciprocal of density and expressed in cm³/g. It is calculated from caliper and basis weight. The bulk is an indication of the relative amount of air in the paper, which in turn affects the optical and strength properties of paper. Decrease in bulk or in other words increase in density makes the sheets smoother, glossier, less opaque, darker, lower in strength, etc.

$$\text{Bulk}\,(\text{cm}^3/\text{g}) = \frac{\text{Thickness}\quad(\mu\text{m})}{\text{Basis weight}\quad(\text{g/m}^2)} \times 1000$$

High bulk is desirable in absorbent papers while lower bulk is preferred for printing papers particularly Bible paper, dictionary paper, etc. The standard procedure of measuring bulk is given in TAPPI T 500.

Wire Side and Felt Side

The side which is in contact with the paper machine wire during paper manufacture is called the wire side. The paper formed on twin wire machines has two wire sides. The felt side is the opposite side of the paper from the wire side (for paper from single-wire machines). The wire side is usually rougher than the felt side because of the pattern caused by wire marks. Generally, paper is open or porous on the wire side and closed or fine textured on the felt side. This is caused by the washing of the fines from the wire side in the manufacturing process. Certain properties differ between wire and felt side such as smoothness, texture and ink absorbency and it is customary to measure these properties on both the sides. This difference of properties on two sides of paper is known as two-sidedness. Highly filled or loaded paper made from short fibre pulp will show higher two-sidedness. Two-sidedness is normally undesirable since it results in a difference in surface properties between the two sides of the sheet, thus affecting the utility of the paper. Two-sidedness will be reduced by beating the pulp. It is generally observed that the mechanical pulp gives more two-sidedness than the chemical pulp.

In the case of paper to be printed on one side only, best results are obtained by printing on the felt side. Postage stamps are printed on wire side and then gummed on felt side, where the smoothness is helpful for attaining an even application of gum. Wire side and topside described above are in reference to single-ply paper. In the case of multi-ply paper, every ply will have wire side and topside. The topside of top-most layers will be felt side and wire side of bottom-most layer is wire side of multi-ply board. Different types of fibres, fillers and chemicals are used in different layers for techno-economical reasons.

Machine- and Cross-Direction

Paper has a definite grain caused by

- the greater orientation of fibres in the direction of travel of the paper machine, and
- the greater strength orientation that results partly from the greater fibre alignment and partly from the greater tension exerted on the paper in this direction during drying.

This grain-direction is known as machine-direction (MD). The cross-direction (CD) is the direction of paper at right angles to the machine direction. The grain of paper must be taken into account in measuring the physical, strength and optical properties of paper. Some of the properties vary with the MD and CD and hence the values are reported in both the directions. The sheets that have all relevant properties same or almost same in both directions are known as "square sheet".

While sheeting the paper, machine- and cross-direction are to be kept in mind and the sheet cutting to be done to suit the end-use requirements.

- All printing papers are to be cut in long grain (the biggest dimension in the grain direction).
- Book papers fold better and the book stays open better if the sheets are cut so that the machine-direction runs up and down the pages.
- Wrap-around labels for metal cans and bottles are to be cut with the machine-direction vertical to obtain greater flexibility about the can.

The sheet is said to be in long grain if the larger dimension is parallel to grain-direction (MD) and in short grain if the larger dimension is parallel to cross-direction.

There is no sure way to determine the MD or CD of a sheet but there is one crude method, the procedure of which is as follows.

1. Cut a strip of about 1" wide and 2" long paper and moist it.
2. Put this moist sheet on a smooth surface or hand. As the sheet will dry, it will curl.
3. The direction of curl is the CD as the paper contracts more in CD than in MD while drying.

Curl

Paper curl can be defined as a systematic deviation of a sheet from a flat form. It results from the release of stresses that are introduced into the sheet during manufacture and subsequent use.

There are three basic types of curls: mechanical curl, structural curl, and moisture curl. Mechanical curl develops when one side of the paper is stretched beyond its elastic limits. The paper will take on a permanent curl if it is allowed to remain wound in a roll for any considerable amount of time. This effect will be more pronounced in the inner part of the roll. This phenomenon occurs because the outer side of the sheet, which is under tension, undergoes a permanent stretching. This type of curl is often called mechanical curl.

Structural curl is caused by two-sidedness in the sheet that is a difference in the level of fines, fillers, fibre area density or fibre orientation through the sheet thickness. Many curls are structural curls in which strains are built into the sheet in the drying process. If the strain is greater on one side than the other, the paper will curl.

Moisture curl occurs when the paper is required to come to equilibrium with new atmospheric moisture conditions and the two sides of the paper react differently, producing a finishing, or uneven absorption of surface-size by the two sides of the sheet. Paper coated on one side only is an extreme example of paper whose sides show very different moisture absorption technologies. Coated one side paper tends to curl towards the coated side when exposed to high relative humidity because the coated side is relatively inert to water vapour and remains dimensionally stable. If the two sides of a sheet have different stresses established in them during drying, they will expand to different degrees when exposed to high relative humidity, thus producing a moisture curl.

The thin papers have less resistance to curl than thick papers. Papers made from long, highly beaten fibres are more likely to curl than are papers made from short, free fibres because greater tensions and unrelieved stresses are set up in the former case. Moist papers tend to curl proportionately less than do papers that are very dry. Fillers decrease the curl because they reduce the overall cohesiveness of the paper.

This paper curl has been a persistent quality issue and is increasingly important for paper grades that are being subjected to high-speed printing, xerography and high-precision converting processes.

The standard procedure for curl measurement is explained in TAPPI T 466 and T 520.

Formation

Formation is an indicator of how uniformly the fibres and fillers are distributed in the sheet. It is thus a physical property of the paper, although it is customarily measured by the degree of uniformity of light transmission throughout the paper. It is judged visually and subjectively by looking through the sheet at a uniform light source. A paper is said to have a uniform formation if the texture is similar to ground glass when viewed in transmitted light. The formation is said to be poor or wild if the fibres are irregularly distributed, giving the sheet a mottled or cloudy appearance in transmitted light. A good formation indicates uniform fibre and filler distribution in the sheet.

Formation is determined by the following two factors.

1. the intensity or density of the clouds or mottled areas, and

2. their spacing.

Formation plays an important role as most of the paper properties depend on it. It will affect properties like caliper, opacity, strength, etc. It also affects the coating capabilities and printing characteristics of the paper. Good formation is essential in writing and printing papers. Therefore, the coated paper should have a good formation.

Friction

Friction is the resisting force that occurs between the two paper surfaces that are in contact when the surfaces are brought to slide against each other. This property is measured as a coefficient of friction that is given by the ratio of the frictional force to a force acting perpendicular to the two surfaces.

Two components of friction can be measured: static and kinetic friction. Static friction is the force resisting initial motion between the surfaces and kinetic friction is the force resisting motion of the two surfaces sliding against each other when already sliding at a constant speed.

Measurement of the coefficient of friction has applications in packaging where a high coefficient will indicate that containers such as sacks, bags and paperboard containers will resist sliding in unit loads or on packaging lines. This property is also important in printing papers, since a specific coefficient of friction is needed so that individual sheets will slide over each other, otherwise double press feeding may result.

There are two methods of measuring coefficient of friction of paper. One which uses inclined plane is explained in TAPPI T 548 and T 815, and the second which uses horizontal plane is explained in TAPPI T 549 and T 816.

Dimensional Stability

Dimensional stability of paper refers to its ability to maintain constant dimensions in grain and cross directions under the environmental conditions and the stresses applied during its printing and converting.

Cellulose fibres (main constituent of paper) swell in diameter from 15 to 20% from dry conditions to saturation point. Since most of the fibres in paper sheet are aligned in the machine run direction, absorption and deabsorption of moisture by paper causes the change in CD dimension. Such changes in dimension may seriously affect the register in printing processes and interfere with the use of such items as tabulating cards. Uneven dimensional changes cause undesirable cockling and curling. Dimensional changes in paper originate in the swelling and contraction of the individual fibres. It is impossible to be precise about the degree of this swelling because paper-making fibres differ considerably in this property, and because the irregular cross-section of fibres creates difficulty in defining the diameter. The change that occurs in the dimensions of paper with variation in the moisture content is an important consideration in the use of paper. All papers expand with increased moisture content and contract with decreased moisture content, but the rate and extent of changes vary with different papers.

Some of the grades of paper requiring a high degree of dimensional stability are printing papers, recording papers, map papers, template papers, abrasive papers, and wall boards. Dimensional stability of paper can be improved by avoiding the fibre to absorb moisture. Well-sized papers have better dimensional stability.

Compressibility

The compressibility of paper refers to its reduction in thickness when pressure is applied perpendicular to the sheet surface. The compressibility of a paper depends upon its hardness and density. It also depends on the nature of mineral fillers and the flexibility of fibres used. The bulky papers are more prone to compressibility than thin dense papers. It influences the ability of paper to change its surface contour during printing and enables the image area of the plate to make contact with the substrate surface. The ability of a substrate to recover its thickness and surface shape is called its resiliency. The compressibility and resiliency of a paper compensate each other for variation in surface thickness and evenness so that the most intimate contact between printing image area and paper may be achieved. The degree of compressibility of a paper mainly depends on its thickness, moisture content, density, hardness, and surface smoothness. It also depends on the degree of calendering.

The compressibility of a paper is a less relevant factor for offset litho printing because the blanket itself is both compressible and resilient. Similarly, this property is less important for flexible plates used in flexography. However, this property is highly relevant in gravure and letterpress printing. Compressibility is measured as a ratio of roughness under two different standard pressures in a Parker Print Surf tester.

Smoothness

Smoothness is an important surface contour of paper. It influences both the functional and appearance properties of a paper. The surface smoothness of a paper is a major factor influencing the print quality. It is common to say that paper has a "smooth" or a "rough" texture. The smoothness and roughness are the two inverse properties of a paper. Surface smoothness is important for writing paper, where it affects the ease of travel of the pen over the paper surface. Smoothness of the paper will often determine whether or not it can be successfully printed.

Surface smoothness of paper is indirectly governed by fibre morphology and treatment during stock preparation and paper-making. Increased beating of the stock increases the smoothness of a paper. Paper-finishing processes such as calendering and coating are usually employed to improve the surface smoothness.

The surface smoothness is important in obtaining a good ink transfer. Since paper is normally being compressed at the moment of ink transfer, information on the smoothness of paper is incomplete without the knowledge of its compressibility. The Bendtsen or Print Surf testers can be used to measure this property by comparing the rate of air leak for different pressures between the sensing head and the paper. Differences in these rates would be greatest for soft papers.

Porosity

As the paper is composed of a randomly felted layer of fibre, it follows that the structure has a varying degree of porosity. Thus, the ability of fluids, both liquid and gaseous, to penetrate the structure of paper becomes a property that is highly significant to the use of paper. Paper is a highly porous material and contains as much as 70% air. The porosity level of a paper depends on the method of its manufacture. A paper made from long fibres will be more porous than one made from short fibres. Fillers make paper less porous by occupying some of the interfibre spaces. Increased refining, by producing greater interfibre bonding and a more tightly intermeshed fibrous structure, decreases the porosity and results in a denser paper. The surface sizing will decrease porosity by sealing most of the surface voids and some of the pores. Calendering makes the surface of paper very compact and hence decreases its porosity.

Porosity (air permeance) is a measure of the air flow, under standard conditions, which will pass through a given area of a sheet of paper. Paper porosity is tested by forcing air under a specified pressure against the sample and measuring the time required for a given volume of air to escape through the sheet. A high air flow indicates that the interfibre void areas are large and the sheet is referred to as being **open**. A low air flow indicates that the void areas are smaller and the sheet is more **closed**.

High porosity or variable porosity, in light-weight papers can cause feeding problems at the suction cups of converting machines or printing presses when two or more sheets or cut blanks may be picked up. However, low porosity may also cause problems. For example, in laminating or labelling papers this would not allow sufficient penetration of adhesive to enable a strong bond to be formed.

Porosity is a property of direct importance in writing and printing papers since it is a factor in the absorption of inks. An indication of ink penetration and degree of spreading during printing can be obtained from porosity. In the case of unimpregnated electrical insulating papers, porosity is an important property because of its influence on the dielectric constant. Porosity is essential in bag papers to be filled by valve connection where the bags must have a certain porosity to prevent bursting during filling. In cigarette papers, the porosity must be controlled through the use of filler (calcium carbonate) to regulate the burning rate.

In addition to air flow, it also gives an approximate measure of the permeability of paper to other fluids such as oils and solvents. Porosity is very important in filter papers used for the filtering of oils and aqueous fluids.

Absorbency

Paper is basically made up of cellulose fibres which are randomly felted to form a network of fibres. Thus the structure of paper consists of a large number of small cavities or holes, in between the intermeshed fibres, which together form a maze of narrow channels capable of drawing in liquid by the action of surface tension. The rate at which liquid is drawn into the paper will be governed by the bulk of the paper (the number and size of the pores) and the properties of the liquid (its surface tension and viscosity).

Absorbency of paper can be related to both water and oil (ink) absorbency. Bulky (open) papers with large pores may absorb the pigment of an ink along with its vehicle, whereas a coated paper having fine pores may only absorb the vehicle. The absorbency of a paper depends upon the composition of the stock, the degree of refining and calendering, filler content, sizing, final sheet density, and surface coating. The degree of absorption of an ink to its substrate is of utmost importance in printing since the final appearance of a print depends to a large extent on the rate of penetration of the ink.

Ink absorbency is the property that determines at what rate and in what amount the ink penetrates the paper, after the press plate or blanket deposits it. Certainly many printing faults can be traced back to differences in ink absorbency of paper. If the ink absorbency of paper is too low (as in coated papers), the ink will tend to lie on the surface of paper in wet condition (due to slow setting) and may set off on to the under side of the next sheet in the stock. If the rate of ink absorbency of the paper (particularly in uncoated papers) is too high, then there is excessive penetration of ink resulting in loss of density and gloss. The excessive penetration of ink results in the situation at which the opacity of the paper cannot obscure the ink image when viewed from the opposite side of the printed sheet. An image can then be seen from both sides of the paper and this is called show-through. Another serious problem is strike-through, in which the ink penetrates right through the paper wetting or sticking to the sheets below. Hence, it is very essential to match the absorbency of the paper to the rheology of the inks and consequently the type of printing process. The extent of ink penetration can also increase with the increase in printing speed or pressure.

K and N test is the simplest test used to measure the ink absorbency. In this test, a thick layer of a specially formulated ink containing dyed oil is spread on to the paper surface and wiped off after 2 minutes. The intensity of the stain left behind is taken as a measure of ink absorption and is visually assessed using a reflectance spectrometer. A better technique makes use of IGT tester, in which measurement can be made under proper press conditions.

Water absorbency of paper is more significant to the paper manufacturer than the printer. The degree of sizing of paper is related to water absorbency since the function of sizing a paper is primarily to render it resistant to penetration by water. Unsized paper (waterleaf) behaves like blotting paper and cannot be printed on with ink because the ink spreads or feathers.

Controlling of water absorbency of paper is essential in such applications, in which paper is going to be treated with aqueous or water-based materials. These include coating with minerals, application of water-based adhesives in bookbinding and carton manufacture. The paper used in offset lithography must be sized to provide adequate water resistance property. Sized papers are most suitable for printing with water-based inks. The Cobb test provides a simple and effective means of measuring the surface water absorbency of paper.

STRENGTH PROPERTIES

The most important strength properties of paper are stiffness, tensile strength, burst strength, tear resistance, folding endurance and surface strength.

Stiffness

Stiffness is the measure of force required to bend a paper through a specified angle. The high bending stiffness of paper and board is attributed to the high specific stiffness of the cellulose fibre which in turn imparts stiffness to the fibre network. The main criteria which govern stiffness are the nature of the fibre, its orientation, dimensions and bulk. The nature of the fibre includes factors like fibre length, cell wall thickness and lignin content. The moisture content of the paper also affects its stiffness. The relative stiffness decreases with the increasing moisture content, changing between 5 and 10% for each 1% change in moisture content.

Papers made from highly beaten stocks (bond and glassine) have higher stiffness than do papers made from lightly beaten stocks (towelling and filter papers). Papers made from short-fibred pulps are generally stiffer than are those made from long-fibred pulps. Papers made from chemical wood pulp generally have a higher stiffness.

Stiffness is an important property for boxboards, corrugating medium and to certain extent for printing papers also. A flimsy paper can cause feeding and delivery problems in larger sheet presses. On web presses, a uniform stiffness across the sheet is required to ensure good registration. A sheet that is too stiff will cause problems in copier machines where it must traverse over, under, and around feed rollers. Bond papers also require certain stiffness to be flat in typewriters, etc. Post-printing processes such as folding require an appropriate degree of stiffness to ensure good convertibility.

The Taber stiffness measures the bending moment of a vertically clamped 38 mm (1.5 inch) wide paper specimen at 15° from its centre line (TAPPI Standard T 489). Taber stiffness is important for papers used in wrapping, structural uses and printing. TAPPI Standard T 45 describes the Gurley tester for measuring stiffness.

Tensile Strength

Tensile strength is a good indicator of the ability of paper to resist stresses imposed by web printing and converting operations, and end-use demands for virtually all packaging applications, particularly bags and sacks.

It is defined as the tensile force per unit width required for breaking a strip of paper or paperboard and expressed in kN/m. The tensile strength of paper can also be represented in terms of tensile index which is the tensile strength divided by its grammage.

The tensile strength is measured on paper strips 20 cm (7.9 inch) long by 15–25 mm (0.6–1.0 inch) wide using a constant rate of elongation according to TAPPI Standard T 494. It is measured both in machine-direction and cross-direction using a suitable tensile tester.

Tensile strength of paper is always greater in the machine-direction than in the cross-direction because of the greater alignment of fibres in the machine direction. The ratio between machine- and cross-direction tensile strength is an indication of the squareness of the sheet.

Tensile strength is an indicative of fibre strength, fibre bonding and fibre length. Tensile strength failure will occur if bonding between the fibres is weak or if the fibres themselves fail. Extensive refining leads to the formation of a paper with good tensile strength.

Burst Strength

The burst strength is defined as the maximum hydrostatic pressure applied at right angles to the paper surface before it bursts (expressed in kPa). The burst strength of paper can also be represented in terms of burst index, which is the burst strength divided by its basis weight, and is expressed in $kPa\ m^2g^{-1}$. It provides a useful means of evaluating pulps and packaging papers and boards. It represents the general strength characteristics of paper and board. The two main factors responsible for bursting strength are fibre length and interfibre bonding. An increase in fibre length increases the bursting strength. As the interfibre bonding increases, the burst strength rises to a maximum and then decreases. The phenomenon is often observed in beating cycle of a pulp. In the beating process, the increased interfibre bonding is accompained by reduced fibre length, which adds to the reduction in bursting strength. Hence, excessive beating leads to a decrease in bursting strength. Bursting strength is a test of the weakest part of the paper and hence it is affected by formation.

In Mullen test, the paper is ruptured by a small circular (30.5 mm in diameter or 1.20 inch) rubber diaphragm pushed against the paper by glycerine. The maximum pressure before rupture is measured (TAPPI Standard T 403).

Tear Resistance

The tearing resistance is a measure of the work done when a test specimen of paper is torn through a specified distance. Based on the total work done, the force required to propagate the tear can be calculated. It is measured in both machine-direction and cross-direction and is expressed in mN (milli Newton). The procedural standards are explained in TAPPI Standard T 414.

Tearing resistance indicates the behaviour of paper in various end-use situations, such as evaluating web runnability, controlling the quality of newsprint and characterizing the toughness of packaging papers, where the ability to absorb shocks is essential. Tear is a measure of the ability of the packaging material to withstand rough handling. It is also, in conjunction with tensile strength, an important property relating to the runnability and performance of various converting operations, particularly in bag making.

Fibre length and interfibre bonding are the two both important factors in tearing strength. The fact that longer fibres improve tear strength is well recognized. The longer fibres tend to distribute the stress over a greater area, over more fibres and more bonds, while short fibres allow the stress to be concentrated in a smaller area. Thus tear strength of a paper decreases with refining. Paper with good tear strength can be produced by using long-fibred pulps.

Folding Endurance

Folding endurance is the paper's capability of withstanding multiple folds before it breaks. It is defined as the number of double folds that a strip of 15 mm wide and 100 mm length can withstand under a specified load before it breaks. It measures the strength and flexibility of paper. Long and flexible fibres provide high folding endurance. It is a very complex property relating to the fibre properties such as length, width and cell wall thickness as well as the formation properties of sheet (like degree of hydrogen bonding and sheet homogeneity). Folding endurance has been found useful in measuring the deterioration of paper upon ageing.

It is particularly important for printing grades, where the paper is subjected to multiple folds as in books, maps, or pamphlets. Fold test is also important for carton, boxboards, ammonia print paper, coverpaper, etc. High folding endurance is a requirement in bond, ledger, currency, map, blueprint and record papers. Currency paper has the highest folding endurance (>2000).

The procedural standards for measuring folding endurance using MIT tester are explained in TAPPI T 511.

Surface Strength

The surface strength of paper is the property that enables its surface to withstand the perpendicularly applied forces, without picking or rupturing, during the splitting of an ink film between the printed sheet and the inked plate or blanket.

It is one of the important characteristics that influence both printability and runnability of a paper. On a letterpress or litho machine, only about half of the ink carried on the blanket is actually transferred to the paper at each impression. When an ink film splits during printing, the rupture must take place within the body of the ink film, between the printing surface and the substrate surface, leaving some ink on each. The tack of an ink usually resists this film splitting. During the splitting of an ink film, the paper is subjected to forces acting perpendicular to its surface and opposite in direction to those causing its compression. These forces act so as to disrupt the surface of the paper and to pull the internal structure apart. The magnitudes of these forces vary considerably with the ink tack and printing process. Therefore, the paper must have sufficiently strong surface strength to resist loss of coating (picking) and surface fibres (fluffing). Picking and fluffing causes serious printability and runnability problems.

In coated papers, picking normally occurs due to the poor adhesion of the coating to the body of the paper. As a result, small pieces of coatings are slowly removed from the paper surface during printing. These small pieces of coatings may stick to the printing plate or blanket, and after a few impressions may themselves start to print. However, because they have a certain thickness, they present a slightly raised area to the sheet and print as small solid areas surrounded by a narrow halo of white paper. These characteristic areas on printed sheets are known as secondary picks or hickies and they increase in number with the degree of picking as print run progresses. In addition to this, the blanket requires periodic cleaning and resulting in down time and reduced runnability.

Fluffing occurs when weakly bonded fibres are pulled away from uncoated paper surfaces. This pulled off loose material may adhere to the printing surface, so causing blemishes on successive prints. Fluffing can be controlled by surface sizing of papers. The weakness in the body of the paper may also result in splitting, which usually occurs both with coated and uncoated papers. This may lead to the delamination of paper, the separation of multi-ply papers and boards, and shock-induced failures during converting processes.

The Dennison wax test is used for measuring the surface strength of uncoated papers or boards. For coated stocks, IGT test (which was developed by the Instituut Voor Grafische Techniek (IGT) in Amsterdam) is applicable. IGT is a measurement of the surface strength of the paper. Tacky ink is applied to sample of the paper at an increasing speed. As the speed increases, the peeling force applied to the paper also increases, and the speed at which the fibres begin to be pulled from the sheet is recorded as the IGT. A high IGT (>300) indicates a strong surface strength suitable for demanding offset applications.

OPTICAL PROPERTIES

The visual appearance of a paper depends on its optical properties including gloss, opacity, brightness, whiteness, colour and fluorescence. The important factors which determine the optical properties of the finished paper include the type of pulp used, the extent of bleaching, presence of fillers or surface coatings, presence of dyes or coloured pigments, method of stock preparation and sheet formation, and the method of finishing process. For instance, increased bleaching reduces colour and opacity, but improves brightness.

Gloss

The gloss is a measure of the surface-reflecting properties of a paper. The gloss of a paper depends on the amount of specular reflection of light (which is reflected at an equal and opposite angle) from its surface. It can be measured by comparing this amount of light with that reflected from a standard glossy surface (such as a polished black glass). The amount of specular reflection or gloss is dependent upon the optical smoothness of the surface. The gloss of a paper can be greatly improved by employing special finishing methods during its manufacture. Finishing processes like cast-coating and glazing give paper sheets with a highly polished surface.

In the case of uncoated papers, the extent of specular refection of light observed at its surface is very less. Due to the unevenness of the surface, the light is not reflected in a regular way but is scattered in all directions. This is called diffuse reflection. The gloss is the total light leaving the surface from the specular and diffuse components. The uncoated papers usually show lowest gloss due to their matt finish that exhibits the lowest specular and highest diffuse reflectance components. The cast-coated, varnished or laminated papers show the highest gloss. The coated and calendered or supercalendered papers usually show the medium gloss.

The level of gloss desired is very dependent on the end use of the paper. High-gloss papers are particularly used in magazines. The papers used in multicolour printing should have a high gloss to give high intensity or colour saturation.

Generally, gloss of unprinted sheet/board is measured at 75° (except for cast-coated papers). Printed and varnished surfaces are measured at 60°. The standard procedures are laid out in TAPPI T 480.

Brightness

The scattering of light from the surface is the most important optical property responsible for the brightness of paper. The brightness of paper is largely governed by the light scattering properties derived from cellulose fibres, fillers, coatings and also fluorescent optical brighteners.

The brightness of paper can be improved in many ways. Extensive bleaching of pulps reduces the lignin content to a minimum and hence improves the brightness. Addition of fillers with high brightness during the manufacture can improve the brightness of paper. Additives such as fluorescent optical brighteners improve the brightness of paper to a greater extent. These materials convert the invisible UV light to blue light and offset the yellowing effect of lignin.

Brightness is defined as the percentage of blue light reflected from a sample measured at an effective wavelength of 457 nm.

Cellulose pulp has a natural creamy hue, even when bleached to a very high brightness. This yellowness is due to the presence of impurity (lignin) which absorbs light in the blue region of the spectrum. Since the greatest changes in brightness are therefore in the blue region, a single number measurement has been chosen to quantify brightness for quality assurance purposes. The relevant blue reflectance region of the spectrum has been standardized in TAPPI T 452 and ISO 2469, and is defined such that the peak reflectance wavelength occurs at 457 nm.

Whiteness

Whiteness is the extent to which paper diffusely reflects light of all wavelengths throughout the visible spectrum, i.e., the magnitude and uniformity of spectral reflectance measured as the per cent light reflectance for the whole wavelength range. A uniform or perfect white is neutral in tint, reflecting all colours of the spectrum equally. The whiteness of a paper can be properly specified by its red, green and blue reflectance values. The measurement is usually made using a colorimeter having CIE X, Y and Z filters which relate to the tristimulus functions of the cones in the eye. Alternatively, a single number representation of whiteness, called whiteness index, has been developed, since it is easier to understand and communicate. According to the ASTM standard test method E313, the whiteness index (WI) is calculated by the formula:

$$WI = 4B - 3G$$

where, B and G are the blue and green reflectance values respectively, calibrated as 100% against an absolute white. Thus, for a perfect white W = 100%, B = 100% and G = 100%.

The procedural standards for the measurement of whiteness are explained in ISO 11475.

Colour

Colour is the measure of the hue or chroma of light reflected from the surface of paper. The colour of paper, like of other materials, depends in a complicated way on the characteristics of the observer and a number of physical factors such as the spectral energy distribution of the illuminant, the geometry of illuminating and viewing, the nature and extent of the surround, and the optical characteristics of the paper itself.

Colour is related to perception and therefore measured or specified in terms of colour space. A commonly used system is the CIE L, a, b system. This is based on the idea of colour opposites.

L denotes luminosity or degree of lightness and varies from 0 for perfect black to 100 for perfect white. *a* measures redness when positive and greenness when negative. *b* measures yellowness when positive and blueness when negative.

Fluorescence

Fluorescence measures the amount of optical brightening agent present in the paper. Optical brightening agent absorbs UV light and re-emits it as visible blue light. Under lighting with a UV component, this makes the paper appear more blue and brighter.

Thus the total amount of visible light reflected from the surface is increased which in turn increases the brightness. The increase in brightness observed depends on the amount of optical brightener added and the intensity of the exciting UV source. The addition of fluorescent additives increases the brightness only up to certain limits. Moreover they are more expensive and hence they must be used carefully.

Other additives in the paper such as fillers (particularly titanium dioxide) also absorb strongly in UV region but do not re-emit in the visible region. This will affect the normal functioning of fluorescent additives and reduce the brightness. All high white-grade papers have high levels of optical brightener. Fluorescence less than 5 indicates the presence of a very little optical brightener.

Opacity

It is the property of a substrate to resist the passage of light. Opacity refers to the degree to which paper will transmit light. A paper which transmits a proportion of light is said to be translucent whereas one transmitting no light is said to be opaque. Opacity is therefore only a relative expression and depends on the extent to which light is absorbed or scattered within a sheet structure, or reflected back instead of being transmitted. Since scattering is caused by multiple reflections and refractions through cellulose fibres as light passes into the body of the paper, the greater the number of fibre–air interfaces the greater is the degree of scattering and hence the opacity. An increase in the opacity of a paper may be gained by

- raising the grammage to increase the number of fibres to obstruct the passage of light,
- increasing the bulk (decreasing the density) to allow more fibre–air interfaces to scatter the light. This is achieved by introducing more bulky short-fibred pulp to the furnish or by decreasing density with less refining and calendering,
- providing additional scattering points with pigment–air interfaces by adding fillers such as titanium dioxide, or
- increasing the amount of light absorbed by coloured matter. Adding dyes or using a lower brightness pulp will do this.

Thus, opacity is the ability of paper to hide or mask a colour or object in back of the sheet. It is also known as printing opacity of paper.

Printing opacity is defined as the ratio of the amount of light reflected from a single sheet of paper with a black backing (R_0) to the amount reflected from a pile of sheets of the paper (R_∞), thick enough to be opaque. The ratio of the two reflectance values is expressed as a percentage. The procedural standards are explained in ISO 2471. Opacity is particularly important for papers which are to be printed on both sides such as in book and magazines (where the print must not show-through).

CHEMICAL PROPERTIES

The chemical properties of paper mainly depends on the factors such as the type of pulp used, method and extent of pulping and bleaching, and the type and amount of non-fibrous additives added. The main chemical properties are acidity and pH of paper, ash content, moisture content, and permanence of paper.

Acidity and pH of Paper

Acidity in paper is derived mainly from the alum used in the internal sizing of paper. It may also be derived from the bleach residues left in the pulp, absorption of acid gases from the atmosphere, the presence of organic acids found in the pulp, or coating materials. Uncoated papers are usually slightly acidic, with a pH range of 4.5–6.0. Most of the coated papers have a pH above 7, due to the alkalinity of the coating.

The acidity of paper may be determined as the amount of water-soluble acidity (or alkalinity) or as the hydrogen ion concentration (pH) of the paper extract. The hydrogen ion concentration (pH) is better indicative of the stability of paper than is the total acidity.

The pH value of paper means the pH of an aqueous extract of the paper, which is prepared by a standard procedure.

The pH of paper can be determined by either cold or hot extraction method. In the cold extraction method, 1 g of paper (either cut or ground) is macerated with a strong rod in 20 ml of distilled water. Then the sample is added with 50 ml of water and left for 1 hour. Finally the pH of the solution is accurately measured by using a pH meter. The same procedure is followed in the hot extraction method, except that the sample is digested at 100°C under reflux for 1 hour instead of keeping at room temperature. The hot extraction method normally gives a pH about 0.5 lower than that from cold extraction method. This is due to the increased hydrolysis of aluminium salts at high temperature. Because of the difference in results obtained with the two methods, the analytical report should always state which method was used.

The pH of paper is one of the factors influencing the rate of drying of a litho ink or letterpress. Generally the effect of pH of paper is significant in the case of inks which dry by oxidation and polymerization. The lower the pH, the longer is the time taken by the ink to dry. However, at pH below 5, the drying may be greatly retarded or even stopped altogether. The problem is more acute when a paper with a low pH is printed in an environment with a high relative humidity of air. The drying time becomes longer at higher moisture contents and lower pH values. A low or high pH paper can cause adverse effects on the printing ink used. Some ink pigments may remain unstable in strongly acidic or alkaline conditions. These conditions in a paper may cause a particular ink colour to fade out of a print.

The acidity of paper is important because of its effect on permanence. The pH is very important for bonds, ledgers and index papers that are intended for permanent records. The pH is also important in converting papers (such as papers for saturating, impregnating or coating) since it is likely to affect the materials used in the converting operations. In ground wood papers, low pH improves the colour and prevents picking on the presses.

Ash Content

The amount of mineral residues left after complete combustion of paper at high temperature represents the ash content of the paper. It is generally expressed as per cent of original test sample and is a measure of filler content (such as clay, calcium carbonate, titanium dioxide) in the paper.

The ash content of paper is determined by igniting a known weight of the paper in a platinum crucible, using an electric muffle furnace at about 900°C. The ignition is carried out until the ash is free from specks of carbon, and this generally requires about 30 to 60 minutes.

The uncoated and unfilled papers usually have 2 to 5% of ash content. Printing papers for letterpress generally have an ash content of 15 to 25% because of the pigments used in filling. Bond and writing papers usually have an ash content of 2 to 6%. The wrapping papers and container boards should have a low ash content since mineral matter tends to reduce the strength of paper. Filter papers used for analytical work must be ash free. High ash content is also undesirable in photographic papers and electrical insulation papers.

Moisture Content

The moisture content of paper is defined as the loss of weight of a sample after oven drying at a constant temperature of 105°C. The moisture content of a paper is normally measured by weighing a sample on an electronic balance and then drying it in a hot air oven at 105°C until it reaches a constant weight. The moisture content is expressed as the percentage loss in the original weight of the sample. Alternatively, samples may be dried on a sensitive enclosed balance using an infrared drier. Modern moisture meters (Moistex MX 5000) use microwave technology based on the concept that polar water molecules absorb microwaves. The amount of microwave power lost on passing through the paper is a measure of its moisture content.

Moisture content in paper varies from 2–12% depending on relative humidity of air, type of pulp used, degree of refining, and chemical used. Most physical properties of paper undergo change as a result of variations in moisture content. Water has the effect of plasticizing the cellulose fibre, and of relaxing and weakening the interfibre bonding. The amount of water plays an important role in calendering, printing and converting processes. Moisture control is also significant to the economic aspect of paper-making. Poor moisture control can adversely affect many paper properties. Moisture influences most paper properties such as smoothness, strength properties (burst, tear, tensile, stiffness and fold), ink receptivity, and dimensional stability in varying degrees. Moisture content of a particular paper depends on drying conditions during manufacture, as well as conditions of storage and use. Maintaining a uniform moisture profiles is an essential requirement in ensuring good printing and converting performance.

Typical moisture values (accepted trade tolerance ±10%)

Paper grade	%
Tissue	2–7
Office/business paper	4–4.5
Printing paper	6–7
Newsprint	7.5–9.5
Marketing wood pulp	10

Permanence of Paper

Permanence is the degree to which paper resists deterioration over time. Permanent paper can resist large chemical and physical changes over and extended time (several hundred years). These papers are generally acid-free with alkaline reserve and a reasonably high initial strength. Papers containing pure cellulose fibre are more permanent. Permanency is desirable in currency, bond and record papers.

PRINTING PROPERTIES

The two important groups of properties that affect the printing of paper by any process are runnability and printability. These are essential for an efficient production of good quality printing in the pressroom.

Runnability

Runnability of paper includes those properties that permit smooth feeding of the paper through the press. They include properties that permit passage of the sheets or web through the press at the desired speed without causing any trouble to the press operation, or damage to the paper or its surface. Factors that affect the runnability of paper include flatness, dimensional stability, surface strength, tensile, burst and tear strength, moisture content and pH. Some properties are more relevant to web-fed than sheet-fed papers and vice versa. Once the runnability is satisfactory, the printability of the paper determines the quality of the prints that can be obtained in a particular printing process.

Runnability is more of a problem in offset than in letterpress or gravure because of the overall contact of the paper with the blanket during impression, and the use of water and tacky inks. The mechanical conditions of paper affect the runnability. Paper should be free from holes, wrinkles, scraps, turned-over corners, stuck spots or edges and foreign matters. The paper rolls should be evenly and tightly wound with smooth even edges and a minimum of splices.

Printability

The printability of a paper may be regarded as its ability to consistently reproduce images to a standard quality or uniformity. This is influenced by the printing process and can be evaluated in terms of dot reproduction, dot gain, print gloss, hue shift and print uniformity. There are many paper properties that will influence these factors including surface smoothness, ink absorption, gloss, opacity and colour.

Printability can be judged from the results of individual physical and optical properties of the paper, such as gloss, smoothness, and ink receptivity, but to interpret these results in terms of printability requires considerable experience with the particular grade of paper in question. Furthermore, because so many factors combine to influence printability, it is possible to make serious misjudgments from the analysis of individual tests, and this is the reason why printability is customarily measured by an actual printing of the paper.

Most letterpress testing for smoothness, ink receptivity, and coverage can be done on special test proof presses. Printability testers can be used to make reasonably accurate predictions of picking, ink coverage, receptivity, and with new models, even trapping has been predicted with some degree of success. These are special printability testers for gravure.

For lithography, however, printability testers are far from having the reliability desired mainly because the effects of dampening have been difficult to stimulate. Therefore, most printability testing is done on an offset press.

Print quality A fundamental concept in evaluating paper for printability is print quality. Print quality is defined as the aggregate effect of the various appearance characteristics of printed matter. It is a measure of the degree to which the original copy is reproduced in printing. A comprehensive bibliography on print quality and a classification of print quality factors have been prepared by TAPPI. Both solid prints and half-tones can be evaluated using digital labelled picture points and that are processed in terms of the areas, shapes, and optical density.

The print quality of the final printed paper is determined by the printability of the paper, the ink, and the method of operating the press. The factors affecting the quality of printing are the amount of contrast between the printed and unprinted areas, finish of the printed and unprinted areas, colour, gloss, the uniformity and smoothness of solid and half-tone areas, tone value of half-tones, sharpness, legibility and clarity of detail, and show-through. Print quality is subjective and is usually measured by a visual examination of the print. But methods for measuring objectively the tone reproduction of half-tone prints and the uniformity of solid prints have been suggested.

REVIEW QUESTIONS

1. Describe the structure of wood fibres.
2. What are the main stages involved in paper manufacture?
3. What is the main objective of pulp-making process? How is it achieved?
4. Discuss the principle involved in the following pulp-making processes.
   - i. Mechanical process
   - ii. Chemical process
   - iii. Combined chemical and mechanical process
5. What are the disadvantages of mechanical pulping process? In what way is it better than chemical pulping process?
6. Discuss the method of preparation of:
   - i. Refiner mechanical pulp (RMP)
   - ii. Thermo-mechanical pulp (TMP)
7. Give the principle involved in chemical pulping process. What are the advantages of chemical process over mechanical process?
8. Write short notes on:
   - i. Kraft (sulphate) process
   - ii. Sulphite process
9. Describe the method of screening and cleaning of wood pulp.
10. What is the purpose of bleaching of wood pulp? Describe the multi-stage bleaching used in the case of sulphate wood pulp.
11. Explain the mechanism of action of hydrogen peroxide as a bleaching agent.
12. Discuss the importance of refining process.
13. Explain the effect of refining on paper properties.
14. Explain the internal sizing of paper using rosin.
15. What are the disadvantages of acid-rosin sizing?
16. What are fillers? Discuss their effects on the paper properties.
17. What are the adverse effects of fillers?

18. Give the importance of titanium dioxide and calcium carbonate as fillers.

19. Discuss the importance of colouring materials as pulp additives.

20. Name the main parts of a paper-making machine. Describe the steps involved in paper-making process.

21. Discuss the influence of calendering on paper properties.

22. Explain the soft and hard calendering process. What are the advantages of these processes?

23. What is the difference between a paper machine calender and supercalender? Explain the supercalendering process.

24. Discuss on the materials used for the coating of paper. What are the special properties of coated papers?

25. Describe briefly the main methods of paper coating.

26. Write a note on types of papers.

27. What do you mean by basis weight of paper? Mention its importance.

28. Explain the following terms.
    i. Caliper
    ii. Bulk
    iii. Machine cross-direction
    iv. Curl
    v. Formation
    vi. Wire side and felt side
    vii. Coefficient of friction

29. Discuss the types of curls.

30. What do you mean by the term dimensional stability of paper? How can it be improved?

31. What is meant by compressibility of paper? Mention its significance.

32. Explain the factors influencing the surface smoothness of paper.

33. Describe how the following factors influence opacity of paper.
    i. Presence of fillers
    ii. Level of refining

iii. Surface sizing

iv. Calendering

34. Discuss the importance of porosity of paper.

35. Give suitable reasons for the following statements.

   i. The two-sidedness of paper is undesirable.

   ii. The tensile strength and other paper properties vary depending upon the direction of paper.

36. Discuss the printing problems that occur due to differences in ink absorbency of paper.

37. Explain the following strength properties of paper.

   i. Stiffness

   ii. Tensile strength

   iii. Burst strength

   iv. Tear resistance

   v. Folding endurance

   vi. Surface strength

38. Mention the factors that govern the stiffness of paper.

39. Explain the effect of refining on the following paper properties.

   i. Tensile strength

   ii. Stiffness

   iii. Burst strength

40. What is meant by surface strength of paper?

41. Explain the following printing problems with reference to paper:

   i. Picking

   ii. Fluffing

42. Explain the following optical properties of paper:

   i. Gloss

   ii. Opacity

   iii. Brightness

   iv. Whiteness

   v. Colour

   vi. Fluorescence

43. What is paper brightness? How can it be improved?

44. What is paper opacity? What are the different ways by which the paper opacity can be improved?

45. Explain the extent of gloss exhibited by uncoated, coated and calendered papers.

46. What are the factors which influence the chemical properties of papers?

47. What is meant by pH of paper? How is the pH of paper determined?

48. Discuss the significance of pH of paper.

49. What is meant by ash content of paper? Describe the determination of ash content of paper and give the importance of the same.

50. Write a short note on moisture content of paper.

51. What is meant by runnability of paper? Discuss the factors influencing the runnability of paper.

52. What is meant by printability of paper? What are the factors which influence the printability of paper?

ADDITIONAL READING

Biermann, C.J. (1996). *Hand book of Pulping and Paper-making*, 2nd edn. Academic Press.

Bureau, W.H. (1989). *What the Printer Should Know About Paper*, 2nd edn. Graphic Arts Technical Foundation, USA.

Cosey J.P. (ed.) (1981). *Pulp and Paper (Chemistry and Chemical Technology)*. Vol. III. Wiley International, New York.

Scott W.E. and Abbott, J.C. (1995). *Properties of Paper: An Introduction*, 2nd edn. TAPPI Press, Atlanta, Georgia.

10 ADHESIVES

INTRODUCTION

Adhesive materials are used primarily in the wood, paper and packaging industries. Many products of the printing and packaging industries, including books, cartons, boxes, calendars and showcards, essentially require adhesives in their production. Today, a wide range of synthetic adhesives is commercially available. The selection of the right adhesive suitable for a particular job is generally not an easy task. However, better selection is possible if one has a basic knowledge of why things stick together.

An adhesive is a compound that adheres or bonds two items together. Adhesives may come from either natural or synthetic sources. Some modern adhesives are extremely strong, and are becoming increasingly important in modern construction and industry. Modern adhesives have improved flexibility, toughness, curing rate, and temperature and chemical resistance.

An adhesive is a substance capable of holding solid materials together by means of surface attachment. Adhesion is the physical attraction of the surface of one solid material for the surface of another. The solid materials held together by means of an adhesive are known as adherends. The force holding the two surfaces together is called the force of adhesion. The adhesive bond or adhesive joint is the assembly made by joining adherends together by means of an adhesive.

Practical adhesion is the physical strength of an adhesive bond, which primarily depends on the forces of adhesion. But the magnitude of practical adhesion is determined by the physical properties of the adhesive and the adherend, as well as on the nature of the adhesive bond formed.

Several years ago, the concept of the interphase relative to the more generally understood interface was developed. The interphase is an important perception that allows us to understand the essential contributions of the various entities that are located within an adhesive joint.

The area between the adhesive and adherend is referred to as the interphase region. This interphase region is a thin region near the point of adhesives–adherend contact. The interphase region has different chemical and physical characteristics than either the bulk adhesive or the adherend. The nature of the interphase region is a critical factor in determining the properties and quality of an adhesive bond.

Separate from the interphase is the interface, which is contained within the interphase. The interface is the plane of contact between the surface of one material and the surface of the other. It is often useful in describing the surface energetics. It is also at times referred to as a boundary layer. Between the adhesive and adherend, there can be several interfaces composed of layers of different materials. The concept of the interphase can best be described by examining metal and plastic substrates in general.

THEORIES OF ADHESION

Unfortunately, there is no single comprehensive theory to explain why adhesives stick. Certain theories apply better to specific situations than others. It is, therefore, crucial to be aware of the fundamentals behind each theory.

The actual mechanism of adhesive attachment is not yet very explicitly defined. Several theories attempt to describe the phenomenon of adhesion. No single theory explains adhesion in a general, comprehensive way. Some theories are more applicable for certain substrates and applications; other theories are more appropriate for different circumstances. Each theory has been subjected to much study, question, and controversy. However, each contains certain concepts and information that are useful in understanding the basic requirements for a good bond. The most common theories of adhesion are based on adsorption, mechanical interlocking, electrostatics, diffusion, and weak boundary layers.

Adsorption Theory

The adsorption theory states that adhesion results from molecular contact between two materials and the surface forces that develop usually designated as secondary or van der Waals forces. For these forces to develop, the adhesive must make intimate, molecular contact with the substrate surface. The process of establishing continuous contact between an adhesive and the adherend is known as wetting. Wetting can be determined by contact angle measurements. Complete, spontaneous wetting occurs when the contact angle is zero, or the material spreads uniformly over a substrate to form a thin sheet.

Wetting is favoured when surface tension of the substrate, better known as the critical surface energy, (C), is high and the surface tension of the wetting liquid is low. Low-energy polymers, therefore easily wet high-energy substrates such as metal and glass. Conversely, substrates having low surface energies (e.g. polyethylene and fluorocarbons) will not be readily wet by other materials and are useful for applications requiring nonstick, passive surfaces.

After an intimate contact is achieved between adhesive and adherend through wetting, it is believed that permanent adhesion results primarily through forces of molecular attraction. Four general types of chemical bonds are recognized as being involved in adhesion and cohesion:

electrostatic, covalent, metallic primary bonds and van der Waals forces (secondary bonds). In most practical adhesive applications, secondary bonds are the predominant elements that contribute to adhesive strength.

Mechanical Interlocking Theory

It is the oldest theory of adhesion. At one time, adhesion was thought to occur only by the adhesive flowing and filling microcavities on the substrate. When the adhesive then hardens, the substrates are held together mechanically. The surface of a substrate is never truly smooth but consists of a maze of peaks and valleys. According to the mechanical theory of adhesion, in order to function properly, the adhesive must penetrate the cavities on the surface; displace the trapped air at the interface and lock-on mechanically to the substrate.

One way that surface roughness aids in adhesion is by increasing the total contact area between the adhesive or sealant and the adherend. If interfacial or intermolecular attraction is the basis for adhesion, increasing the actual area of contact will increase the total energy of surface interaction by a proportional amount. Thus, the mechanical theory generally teaches that roughening of surfaces is beneficial because it gives teeth to the substrate (mechanical interlocking) and increases the total effective area over which the forces of adhesion can develop. However, roughening is only effective if the adhesive wets the surface well.

The viscosity, as well as the time of contact of the adhesives on the adherend, plays a substantial role in determining how well a mechanically roughened surface is adhered by an adhesive. This theory is more appropriate to the bonding of porous materials like most papers and boards, and wood.

Electrostatic Theory

This theory is based on the differences in the electro-negativities of adhering materials. If two materials with large difference in electronegativities are brought into contact, an electron transfer from the material of lower electronegativity to that of higher electronegativity can occur. This transfer leads to the formation of an electrical double layer at the adhesive–adherend boundary interface. Hence, the electrostatic theory states that electrostatic forces are formed at the adhesive–adherend interface. These forces account for resistance to separation. This theory gathers support from the fact that electrical discharges have been noticed when an adhesive is peeled from a substrate. Electrostatic adhesion is regarded as a dominant factor in biological cell adhesion and particle adhesion.

Diffusion Theory

This theory assumes the mutual solubility of the adherend and adhesive to form a true interphase. The solubility parameter (the square root of the cohesive energy density of a material) provides a measure of the molecular interactions occurring within the material. Thermodynamically, solutions of two materials are most likely to occur when the solubility parameter of the two materials are matched. In other words, the best practical adhesion is obtained when there is a mutual solubility between the adhesive and adherend. Thus, the fundamental concept of the diffusion theory is that adhesion arises through the interdiffusion of molecules in the adhesive and adherend. The diffusion theory is primarily applicable when both the adhesive and adherend are polymeric, having compatible long-chain molecules capable of movement. Solvent or heat welding of thermoplastic substrates is considered to be due to diffusion of molecules. However, this theory is not applicable to substantially different materials, such as polymers or metals.

Weak-Boundary-Layer Theory

According to the weak-boundary-layer theory, when bond failure seems to be at the interface, usually a cohesive rupture of a weak boundary layer is the real event. This theory largely suggests that true interfacial failure seldom occurs. In most cases, joint failure results from a cohesive failure of a weak boundary layer. Weak boundary layers can originate from the adhesive, the adherend, the environment, or a combination of any of the three.

Weak boundary layers can occur on the adhesive or adherend if an impurity concentrates near the bonding surface and forms a weak attachment to the substrate. When failure occurs, it is the weak boundary layer that fails, although failure may seem to occur at the adhesive–adherend interface. In addition to external contamination, examples of weak boundary layers are corrosion or oxide layers on metal surfaces and low-molecular-weight constituents (e.g. release agents, plasticizers) on polymeric surfaces. Weak boundary layers must be removed by physical or chemical means so that there is no weak link in the adhesive joint, which would contribute to premature adhesive failure.

INGREDIENTS OF AN ADHESIVE

The ingredients of the adhesive mixture are usually determined by the need to satisfy certain fabrication properties of the adhesive or properties required in the final joint. The main ingredient is the binder material that provides the adhesive and cohesive strength in the bond. It is usually an organic resin or a synthetic rubber, or an inorganic compound or a natural product. Depending on requirements some other ingredients are also added. They are as follows:

Diluents

It is used as a solvent vehicle for other adhesive ingredients and also to control the viscosity of the adhesive medium. Liquid resins are added to control viscosity.

Hardeners and Catalysts

These are used as curing agents for adhesive system. Hardeners effect the curing by chemically combining with the binder material. The ratio of hardener to binder determines the physical properties of the adhesive and can usually be varied within a small range. Thus, polyamides combine with epoxy resins to produce a cured adhesive.

Catalysts are also employed as curing agents for thermosetting resins to reduce the cure time and increase the cross-linking of the synthetic polymer. Acids, bases, salts, sulphur compounds and peroxides are commonly used as catalysts.

Accelerators, Inhibitors and Retarders

An accelerator is a substance that speeds up curing caused by a catalyst by combining with the binder. An inhibitor arrests the curing reaction entirely whereas a retarder slows it down and prolongs the storage and/or the working life of adhesives.

Modifiers

Some chemically inert ingredients are added to adhesive composition to alter their end use or fabrication properties. They include fillers, extenders, plasticizers, stabilizers and wetting agents.

Fillers improve the working properties, permanence and strength of the adhesive bond. The commonly used fillers are silica, alumina, metal powders, china clay, and asbestos and glass fibres. Some fillers may act as extenders.

Extenders are substances that usually have some adhesive properties and are added as diluents to reduce the cost of the adhesive. The common extenders used are soluble lignin and pulverized partly cured synthetic resins.

Plasticizers are incorporated in a formulation to provide the adhesive bond with flexibility. They may reduce the melt-viscosity of hot-melt adhesives or lower the elastic modulus of a solidified adhesive.

Stabilizers are added to an adhesive to increase its resistance to adverse service conditions such as light, heat, etc.

Wetting agents promote the interfacial contact between adhesive and adherends by improving the wetting and spreading qualities of the adhesive.

TYPES OF ADHESIVES

The development of plastics and elastomers has rapidly advanced the development of adhesives and has given formulators a wide variety of products that can change and improve various properties of adhesives such as flexibility, toughness, curing or setting time, temperature and chemical resistance. Modern adhesives may be classified in a number of ways. In general, they can be better classified based on the composition of the principal components and based on the mechanism of adhesion involved.

Based on the composition of the principal components, they are classified into two types.

Natural Adhesives

Adhesives based on vegetable (natural resin), food (animal hide and skin), and mineral sources (inorganic materials) belong to this type, e.g. vegetable gums, starch, dextrin, shellac asphalt, natural rubber, albumin, casein, sodium silicate, etc.

Synthetic Adhesives

Adhesives based on elastomers, thermoplastic, and thermosetting resin which are synthetically prepared belong to this type, e.g. synthetic rubbers, polystyrene, polyvinyl alcohol, phenol–formaldehyde resin, epoxy resins, etc.

Based on the mechanism of adhesion, they are classified into four types.

Solvent-Responsive Adhesives

These types of adhesives are usually used in the form of solutions, gels, pastes and dispersions in a suitable solvent. In this type of adhesives, the flow of the adhesive during application and adherence during bonding are caused by the volatile liquid carrier. The main ingredients of these adhesives are adhesive bases, volatile liquid carriers, plasticizers, and adhesion augmenting resins.

The commonly used adhesive bases are vegetable gums and starches, proteins, vinyl polymers, rubber, synthetic rubbers, alkyd resins, rosin, shellac, acrylate polymers, cellulose esters and ethers, etc. The adhesive base is dispersed in a solvent known as volatile liquid carrier (such as alcohols, ketones, esters, aliphatic and aromatic hydrocarbons) using some surface-active agent. The plasticizers (like dioctyl phthalate, tricresyl phosphate, etc.) are added to reduce the hardness and brittleness of the adhesive base. Adhesion-augmenting resins (generally low-molecular-weight resins) are added to increase the degree of tackiness and the duration of tackiness during drying of the adhesive.

The typical examples of solvent derivative adhesives are vegetable gums, starches, protein glues, rubber adhesives, cellulose derivative adhesives, natural resin adhesives, vinyl resin adhesives and acrylic resin adhesives. The cold-working glue is obtained by treating animal glue with acetic acid and then heating the mixture. Rubber adhesives are the dispersions of natural rubber, reclaimed rubber or synthetic rubbers in aromatic or chlorinated hydrocarbon solvents. Acrylic resin adhesives are the dispersions of polymers of acrylic or methacrylic esters in aromatic or chlorinated hydrocarbon solvents. Vinyl resin adhesives are the dispersions of polyvinyl acetate or polyvinyl acetals or polyvinyl butyral in solvents such as lower aliphatic alcohols, ketones, esters or aromatic hydrocarbons. Shellac in ethanol, rosin in castor oil, asphalt in aliphatic hydrocarbons are used as natural resin adhesives.

Solvent-responsive adhesives are employed for different purposes. For example, water-soluble glues, gums and starches are used as adhesives in bookbinding, packaging, labels, envelopes, wallpaper, cardboards, etc. Cellulose nitrate is used in shoe making. Rubber adhesives are used to bond flexible materials like leather. Asphalt adhesive is used in laminating paper for waterproof wrappings. Vinyl resin adhesives have found applications in the bonding of paper and wood.

Pressure-Sensitive Adhesives

A pressure-sensitive adhesive, a material that adheres with no more than applied finger pressure, is aggressively and permanently tacky. The essential requirements of such adhesives are that they should be capable of providing instantaneous adhesion when applied using light pressure and should be removable from the surface without leaving any residue. Pressure-sensitive additives are most widely used in the form of adhesive tapes. These tapes are used for number of applications such as in packaging, sealing, stenciling, masking, reinforcing, electrical insulation, medicinal applications, etc. The application governs the choice of tape backing and the adhesive formulation. A transparent backing having a relatively weak adhesive is used for paper mending. A filament-filled backing having an aggressive adhesive is used for packaging applications. Pressure-sensitive adhesives are also obtainable in aerosol form for use in various graphic arts applications. For surgical tapes, cotton cloth is used as tape backing. Tape-backing for electrical industry is made of acetate rayon yarn or fibreglass fabric yarn. Cellophane, cellulose acetate and even thin metal foils have also been used as tape-backing.

The pressure-sensitive adhesives are formulated by making use of materials such as elastomers, tackifying resins, mineral fillers, antioxidants and plasticizers. The most widely and perhaps the earliest elastomer used to formulate pressure-sensitive adhesives is natural rubber. Other elastomers that have been used are butyl rubber, acrylic polymers, polyvinyl ethers, alkyd resins, and silicones. Tackifying resins are added to improve tackiness of the adhesives. Silicone pressure-sensitive adhesives are typically tackified by the addition of silicone gum and silicone resin to the silicone elastomer. Mineral fillers such as ZnO are sometimes added as colouring agents and to reduce the cost of the adhesive. Antioxidants are added to provide prolonged life to the adhesive Plasticizers such as pine tar and methyl ester of hydrogenated rosin are added to improve adhesion, tack, wetting power and flexibility of the adhesive material. They are generally applied by calendering or solvent-spreading method.

In formulating a pressure-sensitive adhesive, a balance of three properties i.e., shear strength, peel strength, and tack, needs to be taken into account. The shear strength of the adhesives is usually measured by hanging a weight on the end piece of tape and measuring the time of failure. Tack is the technical term applied to quantify the sticky feel of the material. In general, the shear strength and the tack of a pressure-sensitive adhesive increase and then go through a maximum as a function of the amount of tackifying resin added. The peel strength usually increases with the amount of tackifying resin.

Pressure-sensitive adhesives are designed for either permanent or removable applications. Examples of permanent applications include safety labels for power equipment, foil tape for automotive interior trim assembly, and sound/vibration damping films. Some high performance permanent pressure-sensitive adhesives exhibit high adhesion values and can support kilograms of weight per square centimetre of contact area, even at elevated temperature. Permanent pressure-sensitive adhesives may be initially removable (for example to recover mislabelled goods) and build adhesion to a permanent bond after several hours or days.

Removable adhesives are designed to form a temporary bond, and ideally can be removed after months or years without leaving residue on the adherend. Removable adhesives are used in applications such as surface protection films, masking tapes, bookmarks and notepapers, price-marking labels, promotional graphics materials, and for skin contact (such as wound care dressings, athletic tape, etc.). Some removable adhesives are designed to repeatedly stick and unstick. They have low adhesion and generally cannot support much weight.

Heat-Sealing Adhesives

They are non-volatile thermoplastic materials that can be heated to melt and then applied as a liquid to adherend. The bond is formed when the adhesive resolidifies. They are also known as "hot-melt" adhesives. These adhesives have become popular for crafts because of their ease of use and the wide range of common materials to which they can adhere. The ease of application as well as the bonding strength developed depends upon the temperature, pressure and time. The important properties exhibited by heat-sealing adhesives are toughness, flexibility, good electrical qualities, and resistance to moisture, solvents, and chemicals.

Heat-sealing adhesives are prepared from thermoplastic polymers that may be compounded with other materials. Ethylene–vinyl acetate copolymers, ethylene–ethyl acrylate copolymers and low density polypropylene are used in the compounded state, while polyesters, polyamides, and polyurethanes are used mostly in the uncompounded state. The most widely used thermoplastic polymer is the ethylene–vinyl acetate copolymer, which is obtainable in a wide range of molecular weights as well as in a variety of compositions. Plasticizers such as polybutenes, phthalates, and tricresyl phosphates are also added to improve the mechanical shock resistance and the thermal properties of the adhesive. Tackifying resins and antioxidants are also be used. Fillers are added to opacify or to modify the flow characteristics of the adhesive, as well as to reduce the cost. Wax is also a very important component which alters the surface characteristics by decreasing both the surface tension and viscosity of the adhesive in the melt.

These adhesives are normally applied using an automatic applicator or a hand-held gun. Problems in using these adhesives are usually associated with the lack of high temperature performance because, being thermoplastic, they tend to creep under load. These problems can be solved by adding some cross-linking agents. Heat-sealing adhesives are widely used in book binding, shoe and furniture manufacture. Box and carton sealing is a very large use area, as is that of corrugated paper manufacture.

Chemically Reactive Adhesives

This type of adhesives undergoes chemical changes during the formation of bond. During bond formation, different types of chemical changes may occur such as cross-linking, condensation, and polymerization.

Chemically reactive adhesives are capable of forming an adhesive bond that is resistant to heat, solvents and chemicals. A small amount of a reagent may bring about cross-linking in the adhesive. Due to this, the solubility and the softening point of the adhesive change drastically. For example, tanning of protein adhesives (such as soyabean proteins and liquid animal proteins with tanning agents like tannins, aluminium and chromium salts, etc.) improves their water-resisting properties. Similarly the vulcanization of rubber adhesives using vulcanizing agents (such as sulphur, zinc oxide) gives greater shear strength, less creep under stress and more durability to the adhesive. In vegetable and animal glues, vulcanized natural rubber, and synthetic rubber adhesives, cross-linking changes take place. Tackifying resins (such as rosin or its esters phenol-formaldehyde resins, terpene polymer resins, etc.) are usually added to the adhesives in which cross-linking changes occur.

Phenol–formaldehyde and urea–formaldehyde resin adhesives, on curing, undergo condensation and cross-linking reactions. Urea–formaldehyde resins are used as hot- or cold-setting adhesives. Acid or acid salts are used as catalysts in them.

Some monomers of resins also undergo polymerization to give an adhesive. For example, monomers of allyl esters, acrylic esters and methacrylic esters are polymerized to give complex polymeric resinous adhesives. Allyl esters, styrene and solution of polymethacrylate in methyl methacrylate are examples of adhesives in which polymerization takes place. They generally have plastic monomers—mixture of monomers or a solution of plastic in monomers. They are usually added with peroxide as a catalyst. These adhesives form true chemical bonds with plastics.

UV-curing adhesives contain monomers that cross-link upon exposure to ultraviolet light to form a polymer. The cross-linking (curing) can happen in less than a second at proper energy levels, so these adhesives can be used in high-speed situations. Acrylic adhesives lend themselves to UV curing quite well, but UV cure versions of silicones, urethanes/acrylic blends and cyanoacrylates are also used.

PHYSICAL FACTORS INFLUENCING ADHESIVE ACTION

The following are some of the important physical factors that have a marked influence on the bond strength developed by an adhesive.

Surface Tension

A liquid adhesive is suitable for an adherend surface only if it can wet the surface properly. The wetting tendency of a liquid adhesive mainly depends on its properties like surface tension and viscosity. The force of attraction between the adhesive and the adherend is maximum when the interfacial tension between the two is minimum.

Porosity and Smoothness of the Adherend Surfaces

The porous materials like paper, wood, leather, etc. usually have a larger number of capillaries, which conduct the more mobile portions of the adhesive into it. As a result, the equilibrium between the solute and solvent in the adhesive is disturbed. In such cases, a weak adhesive bond is developed. But in certain situations, such a phenomenon may be advantageous also. For instance, when sodium silicate (adhesive) is used on a porous surface, the quick removal of water by diffusion through the capillaries causes a rapid development of the bond strength. It has also been observed that smooth and properly planed surfaces are getting bonded together more tenaciously than rough surfaces.

Physical Characteristics of Adhesive Film

The bond strength developed is greatly influenced by the physical characteristics such as tensile strength, shear strength, compressive strength, modulus of elasticity, creep rate, and thermal coefficient of expansion of the adhesive film formed. Generally, the greater the tensile strength, shear strength, and compressive strength of the adhesive film, the greater the strength of the bond formed. The adhesive film formed should have a negligible creep under the stress of structure. It has been found that the film formed by a highly plasticized adhesive is bound to creep under stress more readily than that formed by a more rigid thermosetting adhesive. The relative thermal coefficient of expansion of adhesive film and the adherend surfaces also influence on the extent of bond strength developed. If the adhesive film and the adherend surfaces have quite different thermal coefficient of expansion, then considerable stress is developed on the film under a large fluctuation in temperature. However, the coefficient of expansion of an adhesive can be reduced to the minimum by the addition of finely powdered inorganic substances like aluminium oxide.

Thickness of the Adhesive Film

The highly viscous adhesives usually form thick films with too many voids. In such cases, poor bond strength is developed. Highly viscous adhesives are usually applied under pressure. As a result, the thickness of adhesive film formed is reduced and the bond strength is improved. The addition of a solvent or plasticizer lowers the viscosity of a high-polymer adhesive. It may, however, be pointed out that in case of non-volatile liquid adhesives, the strength of the adhesive film formed is independent of its thickness.

Method of Application of Adhesive

The method used for applying an adhesive also influences the extent of bond strength developed because methods of application are dependent on physical form of the adhesive. Irrespective of the method of application of adhesive, the factors like pressure, temperature, and time have definite effect on the final strength of the bond formed. The temperature and time actually determines whether full curing and drying of the adhesive film have taken place or not. The applied pressure indirectly controls the thickness of the adhesive film.

CHEMICAL FACTORS INFLUENCING ADHESIVE ACTION

The important chemical factors influencing the adhesive action are the following.

Polar Nature of Adhesive

The adhesive molecules containing polar groups are strongly adsorbed by the organic contacting surfaces (like wood, plastics, etc.), resulting in stronger bond strength. On the other hand, non-polar polymeric adhesives provide poor bond strength in such cases. It has been observed that the addition of a polar molecule to a non-polar adhesive increases considerably the adhesive strength of the latter. For instance, the adhesive strength of polyvinyl chloride or polyvinyl acetate can be improved by the addition of maleic acid.

Degree of Polymerization

In the case of resin adhesives, the semi-solid fractions provide better bond strength compared to highly cured and tough fractions. Thus, for maximum bond strength, the degree of polymerization of an adhesive should lie in a range between unpolymerized and fully polymerized fractions. The adhesives prepared from cellulose derivatives provide better strength, if they have undergone a partial degradation to lower molecular weight products.

Side Chains in Adhesive Molecule

The chemical nature, length and complexity of the side chains in the adhesive molecules also play a major role in the development of bond strength. It has been found that in the case of cellulose esters better bond strength is developed, if the fatty acid chains have 6 to 14 carbon atoms. Similarly, the resin adhesive products of phenols with higher aldehydes provide better bond strength than that of condensation products of phenols with lower aldehydes.

Effect of pH

Usually the presence of strong acids and alkalis decreases the bond strength. The effect of acids and alkalis is more on wood surfaces rather than on plastic and metallic surfaces. However, protein glues exhibit better bond strength in the presence of lime which keeps the medium only slightly alkaline.

BONDING PROCESSES BY ADHESIVES

The common general steps involved in the bonding process using adhesives are the following.

Preparation of Adherends

Some of the adherend surfaces require a proper cleaning before applying the adhesive. Metal surfaces require chemical cleaning so as to remove dirt, grease, oil and other contaminants before bonding. The chemical treatments may involve solvent cleaning, vapour degreasing, detergent washing, acid or alkali treatment methods.

Wood or metal adherend surfaces are properly machined to get flat and true plained surfaces that provide a thin adhesive film with better bond strength. Plastic surfaces are slightly roughened by sand blasting which helps in better specific adhesion.

Preparation of Adhesives

The preparation of adhesives can be carried out by dissolving the adhesive material in a suitable solvent, or by melting solid adhesives or thinning of liquid adhesives or by addition of additives such as catalysts, hardeners, fillers, extenders and plasticizers to the adhesive material.

Application of Adhesives

The adhesive should be uniformly applied over the adherend surfaces. This can be carried out by spraying of liquid adhesives, roller coating, brushing of adhesive over the surface, laying pressure-sensitive tapes of adhesive, etc.

In the case of chemically reactive adhesive system, one of its component (such as thermosetting resin) is applied on one surface, while the other component (such as catalyst or reactivating solvent) is applied on other surface.

Assembling of Adhesive-Coated Adherends

In the usage of molten thermosetting adhesives, the adherend surfaces should be assembled immediately because the adhesives dry and develop bond strength rapidly. Thermoplastic or rubber adhesive solutions in organic solvents require air-drying for longer periods. In the case of wood-bonding adhesives, the assembled surfaces are held together in close contact for a certain period. In the case of bonding of metals, the adhesive-coated surfaces are heated in an oven before assembly and the process is called pre-curing.

Application of Pressure and Temperature

When the adhesive applied on the adherends has reached the required consistency, the bonding pressure is applied to the assembled joint. The applied pressure is to make the two surfaces to come in intimate contact and to cause the liquid adhesive to flow uniformly over the adherend surfaces, so that no air bubbles or voids remain in the adhesive film. In the case of pressure-sensitive (fast setting) adhesives, the applied pressure is quickly released. On the other hand, in the case of slow-setting adhesives (like thermosetting resin adhesives), the applied pressure is retained for the required time until the full bond strength is developed.

The assembled joints are usually maintained at the required temperature depending on the type of mechanism involved in the development of full strength of the adhesive film. For instance, cooling is required in the case of molten adhesives, whereas heating is necessary in the case of chemically reactive or drying adhesives.

Conditioning of Joints after Bonding

In the application of thermoplastic adhesives, the joint is hot-pressed for bonding. Such joints are cooled below a certain temperature before the pressure is released in order to develop full bond strength. In the case of thermosetting adhesives, the joints are further conditioned by keeping them for some more time after the release of pressure so that full bond strength is developed. In some cases, internal stresses may also develop in the joints after bonding, which can be minimized by proper conditioning.

REVIEW QUESTIONS

1.   What are adhesives? How are they classified?
2.   Explain the terms:
  i.   Adhesion
  ii.   Adherends
  iii.   Adhesive bond
3.   Describe the following theories of adhesion.
  i.   Mechanical interlocking theory
  ii.   Diffusion theory
  iii.   Adsorption theory
  iv.   Electrostatic theory
4.   What are the requirements of a good adhesive?
5.   Discuss the special features of pressure-sensitive adhesives.
6.   Write short notes on:
  i.   Chemically reactive adhesives
  ii.   Hot-melt adhesives
7.   Discuss the various physical factors that influence the adhesive action of an adhesive.
8.   Explain the different mechanisms of development of adhesive strength.
9.   Discuss briefly the chemical factors which influence the adhesive strength.
10.   Describe the general steps involved in the bonding process.

ADDITIONAL READING

Kroschwitz, J.I. (ed.). (1985). "Adhesive and Bonding, Adhesive Composition." In: *Encyclopedia of Polymer Science and Engineering.* Vol. 1. John Wiley and Sons, New York.

Kroschwitz, J.I. (ed.). (1988). "Pressure Sensitive Adhesives and Products." In: *Encyclopedia of Polymer Science and Engineering* Vol. 3. John Wiley and Sons, New York.

Petrie, E.M. (1999). *Hand book of Adhesives and Sealants.* McGraw-Hill, New York.

Skeist, M. (ed.). (1990). *Hand book of Adhesives,* 3rd edn. Van Nostrand-Reinhold, New York.

Satas, D. (ed.). (1989). *Hand book of Pressure Sensitive Adhesive Technology.* Van Nostrand–Reinhold, New York.

CHEMISTRY OF PHOTOGRAPHY

INTRODUCTION

The printing industry is becoming more and more dependent on the processes of photography. Most of the printing processes involve photography in image converting and platemaking. So some basic knowledge of the chemistry of photography is essential to understand these printing processes.

Photography is a process where films or papers coated with a light-sensitive emulsion containing silver compounds are exposed to light to get the image of the object on the film or paper. After exposure, the film or paper is developed, fixed, washed and finally dried.

The light-sensitive materials fall into two main types. First type includes the normal photographic materials based on the use of silver halides. Second type includes a range of photosensitive materials that are used in transferring a photographic image on to a printing surface in producing letterpress, litho- or gravure plates.

PHOTOGRAPHIC MATERIALS

Base Materials

In photography, the films or papers coated with a light-sensitive emulsion are used. The base of photographic paper is made of a special grade of paper. The first film base used for photography was made up of cellulose nitrate, a polymer derived from cellulose. Later it has been replaced as a film owing to its high inflammability, and to the fact that it tends to discolour and become brittle with age. The material that replaced cellulose nitrate as a photographic base film was cellulose acetate. This was commercially developed during the early years of the twentieth century. Cellulose acetate is produced by the action of acetic acid on cotton or wood pulp.

Most graphic art films today employ a polyester base in place of cellulose acetate base. Polyester is obtained by the condensation polymerization of dimethyl terephthalate with ethylene glycol. Polyester film has exceptional properties when compared with cellulose acetate film. Apart from its good dimensional stability, the polyester film is colourless, transparent, tough, and inert to photographic emulsions and insoluble in most solvents.

Photographic Emulsions

Light-sensitive emulsion used in photography is the dispersion of silver halides in a colloidal gelatin. Most silver halides are photosensitive in nature. The gelatin serves to hold and protect the silver halide, and acts as an adhesive to hold the emulsion to the base material, i.e., film or paper. The silver halides used in photographic emulsions are silver chloride ($AgCl$, white), silver bromide ($AgBr$, pale yellow), and silver iodide (AgI, lemon yellow). AgI has a greater light sensitivity than $AgBr$, and $AgBr$ has a higher light-sensitivity than $AgCl$.

The $AgBr$, or a mixture of $AgBr$ and $AgCl$, or mixture of $AgBr$ and AgI are some examples for more light-sensitive emulsions. The emulsions for enlarging paper must be fast and light-sensitive. Such emulsions usually contain $AgBr$ or a mixture of $AgBr$ and $AgCl$. The $AgCl$ emulsion is least light-sensitive and hence used in the emulsion of photographic paper for contact prints. When a negative is to be made from the original copy, a fast emulsion is essential. The fast negative emulsions contain $AgBr$ or a mixture of $AgBr$ and AgI.

The silver halides are almost insoluble in water and can be prepared simply by mixing solution of a soluble silver salt usually $AgNO_3$ and a soluble alkali halide. For example, silver bromide is precipitated when a solution of KBr is added to a solution of $AgNO_3$. In making photographic emulsions, it is necessary to obtain the silver halide in the form of small particles (or grains) evenly suspended in a solution. So, silver halides are formed chemically in a solution of gelatin. Gelatin swells in cold water but does not dissolve. It dissolves in warm water (at about 35°C) to form a colloidal dispersion. If a silver halide is precipitated in the presence of this dispersion, the particles of gelatin are able to carry the separate grains of the halide in an extremely fine form and so act as a "protective colloid". When a gelatin solution (about 2% solution) is cooled, it sets to a "gel". This easy transition from the "hydrosol" to a "hydrogel" provides a convenient means of setting a photographic emulsion when it is coated to a substrate. Another useful property of gelatin is that it can be hardened by chemicals such as potash alum to reduce its tendency to swell in water and to increase the temperature at which it dissolves in water. The size and distribution of the halide grains have an important influence on the sensitivity of an emulsion. So for a precipitation reaction, the conditions such as the concentration and temperature of a solution and the speed of addition must be carefully controlled.

The silver halide grains in photographic emulsions do not contain discrete molecules of silver halide. Instead, an "ionic lattice" is formed in the grain. The silver and halide ions in a photographic grain are arranged in the form of a cubic lattice, with each silver ion having six halide ion neighbours and each halide ion having six silver ion neighbours.

Preparation of silver halide emulsions The photographic film manufacturers employ complicated procedures to produce silver halide emulsions. A simplified version of these procedures is discussed herewith.

A solution containing 1 to 5% gelatin in water is prepared and required amount of KBr is dissolved into it to give a final solution of 10% KBr in gelatin. An aqueous solution of $AgNO_3$ is prepared in a separate container. These two solutions are separately heated to about 70–90° C. Then the hot $AgNO_3$ solution is slowly poured into the hot KBr-in-gelatin solution. The following reaction occurs:

$$AgNO_3 + KBr \longrightarrow AgBr + KNO_3$$

The emulsions prepared in this way usually have grains of widely varying sizes and are more suitable for continuous-tone films. Graphic art emulsions are prepared by a double-jet technique where halide and $AgNO_3$ are simultaneously added to the gelatin solution. Such emulsions usually have a narrow range of grain sizes.

If silver chloride emulsion is required, KCl is to be used instead of KBr and if a mixture of bromide and iodide emulsion is required, then both KBr and KI are to be used with gelatin solution. The sensitivity of silver halide in the emulsion depends on its particle (or grain) size. If the grain size is small, sensitivity is less and vice versa. If silver nitrate is added rapidly, the silver halide particles formed are very tiny, and the sensitivity of the emulsion is low. If the temperature of the solution is raised before they are mixed, and the silver nitrate solution is added slowly, the silver halide particles formed are somewhat larger. Hence the sensitivity of the emulsion is greater. If the initial size of the grains is very small, they increase in size when the emulsion is allowed to ripen. After crystal growth is completed, the gelatin is flocculated by adjusting the pH or by adding a salt. The precipitated gelatin carries the silver halide with it, leaving behind the excess KBr and KNO_3 in solution. Ultrafiltration is another method used to purify photographic emulsions.

After the emulsion has been ripened, gelatin is added some more to increase the total gelatin concentration up to about 10%. Then the mixture is cooled rapidly when it sets to form a gel. The gel is cut into small shreds and washed thoroughly with cold water to remove the excess KBr and KNO_3. Then the emulsion is melted (at about 50°C) and coated on the cellulose acetate film, or polyester film, or paper.

Graphic art films generally contain 2.5 to 5.0 g of silver per square metre. Sulphur- and gold-containing compounds are added as sensitizers to improve the light-sensitivity of the emulsions.

An alternative method is used to prepare silver halide emulsions with larger grains. In this method, precipitation of the silver halide is carried out in the presence of ammonia at low temperatures. Low-speed emulsions used for positive film or process works are prepared by rapid addition of the silver nitrate at a lower temperature. High-speed emulsions used for negatives are made with a slow addition of the silver nitrate and a long ripening at high temperatures. The low-speed emulsions have a finer and more uniform grain size than high-speed emulsions.

STEPS INVOLVED IN PHOTOGRAPHY

Exposure

When a photographic film is exposed to light, scene images are recorded in the emulsion layers by the action of the photo holes and photoelectrons produced in the grains by the absorption of photons. In the unexposed regions of negative film, there are no photo holes and no photoelectrons. Accordingly, no latent images and no catalytic silver clusters are produced and little amplification occurs in these regions during development. However, in the exposed regions, the photoelectrons react with silver ions to form clusters of silver metal. These clusters of silver function as catalytic centres for amplification during subsequent development. Different theories have explained the latent image formation. However, all these theories begin with the same latent image formation process, i.e., photon absorption by the silver halide crystals to produce electron-hole pairs, and these entire theoretical processes end with the same, i.e., with the formation of a sufficiently large cluster of localized silver atoms. Furthermore, all theories describe the formation of silver atoms. The differences among the various theories are in the descriptions of the mechanistic steps that occur between photon absorption and final formation of a stable silver cluster, i.e., latent image.

Development

When an emulsion is exposed to light, no visible reaction occurs. The typical black and white (i.e., exposed and unexposed) portions of a negative do not appear until the emulsion is placed in the developer bath for the required time. This process is called developing. The reagent used for this purpose is called developer or developing solution. When an exposed emulsion is placed in the developer bath, an oxidation–reduction (redox) reaction occurs. Developing agent is the principal ingredient in a developing solution, which is capable of donating electrons, i.e., a reducing agent. The developing agent is oxidized and the exposed silver ions are reduced to metallic silver. The metallic silver is black in the finely divided state and it forms the black part of the negative.

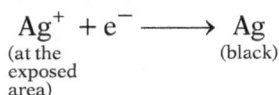

$$\text{Developing agent} \longrightarrow \text{Oxidized developing agent} + ne^-$$

$$\underset{\substack{\text{(at the} \\ \text{exposed} \\ \text{area)}}}{Ag^+} + e^- \longrightarrow \underset{\text{(black)}}{Ag}$$

Composition of a developer solution   A developer solution usually contains the following ingredients.

Developing agents   It is the principal ingredient in a developer solution. Chemically a developing agent should be an active reducing agent. This actually brings about the reduction of silver ions at the exposed area of the film to metallic silver. The successful developing agents in common use are hydroquinone and metol.

A developing agent must be strong enough to reduce only the exposed silver halide grains but not the unexposed grains. For negative films, the electrochemical reduction properties of the developing agent must be properly positioned to provide rapid amplification of exposed grains and much slower spontaneous amplification of unexposed grains. A good developing agent must have the ability to discriminate between exposed and unexposed grains. In other words, an effective developing agent is the one that is selective in its action. Most of the materials meeting these requirements are derivatives of benzene like hydroquinone, pyrogallol, catechol, *p*-aminophenol and metol.

The activity of a developer solution containing hydroquinone can be adjusted by chemical substitutions on the hydroquinone molecule. The substitution of hydrogen atoms on the benzene nucleus with methyl, methoxy, hydroxide or amino groups increases the rate of development. On the other hand, the development activity is reduced by substitution with electron-withdrawing groups such as nitro, sulpho-, cyano-, and carboxyl groups. The activity of a developer solution not only depends on the chemical nature of the developing agents but also on the chemical nature of other components of the developer solution.

Accelerators   The developing agents are only active in alkaline solution. Therefore, a developer solution is usually added with an alkaline material called accelerator. The common accelerators used are NaOH, Na_2CO_3 or $Na_2B_4O_7$ (sodium tetraborate or borax). The accelerators also neutralize the acid formed during developing. The developer with NaOH as accelerator will have a pH of 12.0 and is very active; with Na_2CO_3, pH is about 10.2 and is less active; with $Na_2B_4O_7$, pH is about 9.0 and is least active.

Preservatives   This is added to reduce the effect of oxidation of the developing agent. Thus it helps to keep the solution colourless during mixing and storing. Sodium sulphite (Na_2SO_3) is the common preservative used.

Restrainers   A small amount of KBr is added as a restrainer, which prevents the formation of fog. The bromide ions of the KBr attach themselves to the unexposed silver halide grains. These grains are completely surrounded by bromide ions, so that the developer cannot reach the silver ions and reduce them to metallic silver.

A commonly used all-purpose developer contains metol and hydroquinone as developing agents, sodium carbonate as accelerator, sodium sulphite as preservative and potassium bromide as restrainer.

Types of developers   Two types of developers used in graphic arts are the following.

Infectious developers   These are used in developing high-contrast films. In infectious developers, only hydroquinone is used as the developing agent. Induction period is the time interval between immersion of the film in the developer and the time at which the image begins to appear. In infectious developers, the induction period is long. But when the image begins to appear, its density builds up rapidly. As a result, an image with very high contrast is obtained. The chemicals used in infectious developers are divided into two parts as A and B. Hydroquinone and preservatives are in part A, and alkali and potassium bromide are in part B. The two parts are mixed and used immediately. On working with infectious developers, operators must be careful about the development time, the temperature of the solution, and the amount and kind of replenishment.

Hydroquinone is oxidized to quinone and exposed silver halide is reduced to metallic silver.

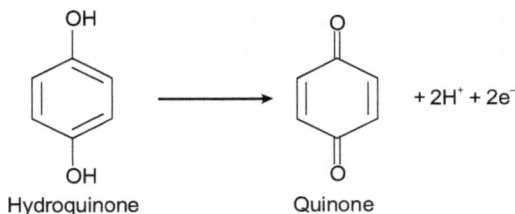

Hydroquinone $\longrightarrow$ Quinone $+ 2H^+ + 2e^-$

$$2\,\text{AgBr} + 2e^- \longrightarrow 2\text{Ag} + 2\text{Br}^-$$
$$\text{(exposed)} \qquad\qquad\qquad \text{(black)}$$

Sodium sulphite used as preservative in the developer can react with quinone to form hydroquinone mono-sulphonate.

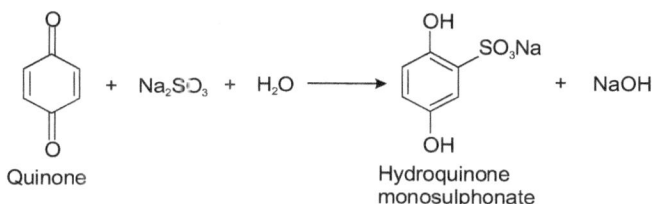

| Quinone | Hydroquinone monosulphonate |

This affects the image-developing reaction and results in an image with lower contrast. Hence, the developer should contain a lower concentration of sodium sulphite. This requirement is often fulfilled by the use of a sulphite buffer instead of sodium sulphite. The commonly used buffer is sodium formaldehyde bisulphate (SFB). In alkaline medium, a small percentage of this compound dissociates to form sodium sulphite and formaldehyde.

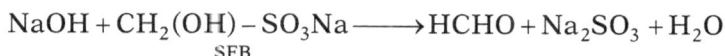

$$\underset{\text{SFB}}{\text{NaOH} + \text{CH}_2(\text{OH}) - \text{SO}_3\text{Na}} \longrightarrow \text{HCHO} + \text{Na}_2\text{SO}_3 + \text{H}_2\text{O}$$

The following are the characteristic features of infectious developers.

- They contain only hydroquinone as the developing agent.
- A sulphite buffer is used to control the concentration of sulphite ions at the minimum level.
- They are sensitive to pH and bromide ion concentration.

Non-infectious developers   These are used in developing camera, scanner and contact films. They produce image densities in proportion to exposure. Non-infectious developers contain both metol and hydroquinone (as developing agents) with high concentration of sodium sulphite (as preservative). Such developers are usually called 'MQ' developers. They are stable and the rate of oxidation is less, so that the development time will be more. All the rapid-access developers are non-infectious. With certain modification in the developer formulation, it is possible to develop film in 60–90 seconds at a developer temperature above 38°C. Rapid-access developer is one such modified developer. It is a low-contrast continuous-tone type developer, containing hydroquinone and metol (as developing agent) and a special restrainer to minimize fogging of film at the high developer temperature. The developer also contains higher concentration of sodium sulphite. The induction period is very short in such developers, and a visible image begins to appear almost as soon as the film enters the developer solution. The rapid-access processing is best suited for films that have a low fogging tendency. It is the commonly used process for the development of contact films. It is also suitable for developing camera films and scanner films.

Fixing

The areas of the negative or positive that have not been reduced to black metallic silver during development still carry a coating of unexposed silver halide (AgBr/AgCl). This coating remains light-sensitive and gradually is reduced to metallic silver on exposure to strong light. The process of removing these unexposed grains is called fixing.

Sodium thiosulphate (hypo) that will dissolve in AgCl or AgBr is used for this purpose in a fixing bath. Ammonium thiosulphate is also used for this purpose. Complex compounds of sodium or ammonium are formed with silver halides and are soluble in water. The reactions involved are as follows:

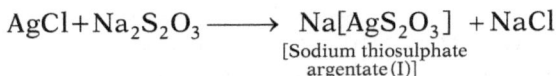

$$AgCl + Na_2S_2O_3 \longrightarrow Na[AgS_2O_3] + NaCl$$
$$\text{[Sodium thiosulphate argentate (I)]}$$

$$AgBr + 2Na_2S_2O_3 \longrightarrow Na_3[Ag(S_2O_3)_2] + NaBr$$
[Sodium dithiosulphate
argentate (II)]

$$AgCl + (NH_4)_2S_2O_3 \longrightarrow NH_4[AgS_2O_3] + NH_4Cl$$
[Ammonium thiosulphate
argentate (I)]

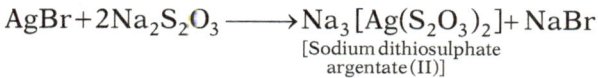

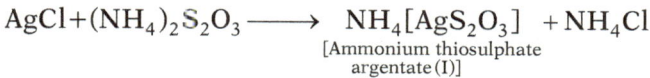

Composition of fixing bath Sodium thiosulphate is the main ingredient in a fixing bath solution. In addition to sodium thiosulphate, fixing bath contains acetic acid, sodium sulphite, potash alum ($K_2SO_4 Al_2(SO_4)_3 \cdot 24H_2O$) and boric acid. Potash alum is a hardener and prevents swelling of gelatin of the emulsion. But it is not stable in neutral or alkaline solutions because of the precipitation of aluminium hydroxide. Therefore, the bath is kept acidic with acetic acid. There is a tendency for developing solution to be brought over into the fixing bath, thus allowing development to continue unless it is checked. So, the acidic bath also neutralizes the alkalinity of the residual developer solution and effectively stops the development process. Unfortunately, sodium thiosulphate is not stable in acid solutions. So, it decomposes to form a white precipitate of sulphur, giving a cloudy bath. The decomposition of sodium thiosulphate can be controlled by the presence of sodium sulphite in the fixing bath. Sodium sulphite chemically reacts with sulphur to form more sodium thiosulphate.

$$Na_2SO_3 + S \longrightarrow Na_2S_2O_3$$

Thus sodium sulphite stabilizes the fixing bath and prevents the formation of a cloudy bath.

Boric acid acts as a buffer and prevents the change of pH of the fixing bath solution. It helps to prevent the precipitation of aluminium compounds that occurs when there is too much variation of pH.

Chrome alum ($K_2SO_4 \cdot Cr_2(SO_4)_3 \cdot 24H_2O$) is also used as a hardener, particularly for processing at high temperatures. The chrome alum is a more effective hardener but is less stable than potash alum. As the fixing process continues, the amount of complex silver compounds formed also increases. When the concentration of silver reaches a critical value (6 g/L in the case of film and 2 g/L in the case of prints), some relatively insoluble compounds are formed that cannot be removed from the emulsion during the washing process.

The pH of a fresh fixing bath solution is about 4.1. The residual developer solution on the film gradually raises the pH of the fixing bath. But the hardener in the bath becomes less effective when the pH rises to about 5.5. At this condition, the bath should be discarded or the pH is lowered by the addition of more acetic acid. Proper cleaning of the films or prints before they reach the fixing bath can prevent the rise in pH of the fixing bath.

Washing

After the fixing process, the gelatin emulsion layer of the negative or positive usually contains the fixing solution dispersed throughout the gelatin. If the hypo (sodium thiosulphate) content of this solution is not removed completely from the gelatin emulsion layer, it will slowly react with the metallic silver in the emulsion to form yellowish brown sulphide.

$$Na_2S_2O_3 + 2Ag \longrightarrow Ag_2S + Na_2SO_3$$

If complex silver compound is present along with hypo in the emulsion, a reaction also occurs resulting in the formation of silver sulphide. This leads to an overall discoloration of the film. To avoid these problems, the film must be washed thoroughly with water so as to completely remove the unwanted materials from the emulsion layer. The washing is usually carried out by placing the films in a tray of water. The process is much more efficient if the films are washed in running water system with water entering at the bottom and leaving at the top of the tray. The use of moderately warm wash water increases the rate of removal of hypo.

PHOTOGRAPHIC REDUCERS

The process of lowering the image density of the original negative is called image reduction. The photographic image reduction is achieved by chemical oxidation of some of the metallic silver to scluble silver salt, and dissolving this away in a suitable solvent like water. The solutions used for this purpose are called photographic reducers. In some cases, the silver salt formed is insoluble in water and hence the reducer solution must contain another chemical that converts the silver salt into a soluble silver compound. For example, a mixture of ferric ammonium sulphate and sulphuric acid is used as a reducer. The reaction involved is as follows:

$$Fe_2(SO_4)_3 \cdot (NH_4)_2SO_4 + 2Ag \longrightarrow Ag_2SO_4 + 2FeSO_4 + (NH_4)_2SO_4$$
Ferric ammonium sulphate

The silver sulphate formed is readily soluble in water and can be easily washed away.

The commonly used Farmer's reducers contain a mixture of potassium ferricyanide and sodium thiosulphate. Potassium ferricyanide oxidizes metallic silver to silver ferrocyanide and reduces itself to potassium ferrocyanide. Then the sodium thiosulphate reacts with insoluble silver ferrocyanide to convert into soluble complex ions such as $AgS_2O_3^-$ and $Ag(S_2O_3)_2^{3-}$.

$$4K_3[Fe(CN)_6] - 4Ag \longrightarrow Ag_4[Fe(CN)_6] + 3K_4[Fe(CN)_6]$$
(Silver ferrocyanide)

$$Ag_4[Fe(CN)_6] + 4Na_2S_2O_3 \longrightarrow 4Na[Ag(S_2O_3)] + Na_4[Fe(CN)_6]$$

$$Ag_4[Fe(CN)_6] + 8Na_2S_2O_3 \longrightarrow 4Na_3[Ag(S_2O_3)_2] + Na_4[Fe(CN)_6]$$

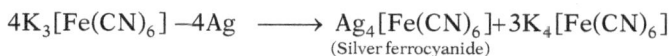

A mixture of sulphuric acid with cerric sulphate is another reducer. In the presence of sulphuric acid, cerric ions oxidize black metallic silver to silver ions and getting reduced to cerrous ions. The ionic equation for the reaction is

$$Ce^{3+} + Ag \longrightarrow Ag^+ + Ce^{2+}$$

Farmer's reducer lowers all densities by an equal amount and so proportionally has the greater effect on the lower densities. The subtractive reducers of this type are most suitable for those negatives which have been correctly developed to the right contrast but have a high density due to gross over-exposure. On the other hand, proportional reducers lower the density of a coating with an accompanying reduction in contrast. It may be used when a negative has been correctly exposed but over-developed. In all cases, reduction involves the chemical oxidation of some of the metallic silver in the image followed by or concurrent with fixation. Ferricyanide is a commonly used oxidant and thiosulphate can be used as the fixing agent.

PHOTOGRAPHIC INTENSIFIERS

The process of increasing the image density of an original negative is called image intensification. The chemicals used for this purpose are known as image intensifiers. An image is intensified by adding a metal, e.g. mercury, chromium, silver or copper, to the image silver. If a given negative had sufficiently detailed information of an original scene but was underdeveloped during processing, then intensification may be useful.

Intensification with mercury involves the treatment of the negative with a mercuric halide, which oxidizes elemental silver and partly converts the silver image to mixture of silver halide and mercury salt. The bleached image is then darkened by a treatment with a developer. During this step, both the silver and mercury ions are reduced electrochemically to their metallic forms, and thereby the image density is intensified.

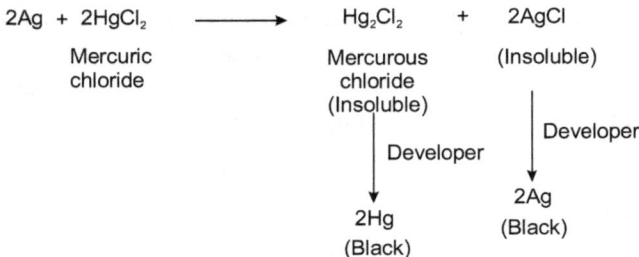

$$2Ag + 2HgCl_2 \longrightarrow Hg_2Cl_2 + 2AgCl$$

Mercuric chloride	Mercurous chloride (Insoluble)	(Insoluble)

Developer (under Hg₂Cl₂) → 2Hg (Black)

Developer (under 2AgCl) → 2Ag (Black)

In an alternative method, single-step mercury intensification is carried out by using a solution containing mercuric iodide and sodium sulphite. Mercury intensifiers may be applied several times in succession if necessary and a weak image may be strengthened to any extent.

REVIEW QUESTIONS

1. What is photography? Name the different steps involved in photography.

2. Give the special features of photographic emulsions.

3. Describe the preparation of silver halide emulsions.

4. Give the function of gelatin in photographic emulsions.

5. Discuss the mechanism involved in the latent-image formation.

6. Describe the mechanism involved in the developing process.

7. Name the ingredients of a photographic developer? Explain their functions with one example for each.

8. Discuss the types of developers.

9. Explain the function of a developing agent. Name the chemical materials that are widely used as developing agents.

10. Give suitable reasons:
    i. A photographic developer should contain a lower concentration of sodium sulphite.
    ii. In the case of infectious developers, a sulphite buffer is used instead of sodium sulphite.

11. What are the characteristic features of infectious developers?

12. Write a short note on non-infectious developers.

13. What is meant by fixing process? Explain the mechanism involved in the process.

14. Give the function of sodium thiosulphate (hypo) in a fixing bath

15. Name the ingredients of a fixing bath. Give the specific function of each ingredient.

16. What is the purpose of the washing process?

17. What are photographic reducers? Explain the mechanism of their action with a suitable example.

18. Write a short note on intensifiers.

ADDITIONAL READING

Duffin, G.F. (1966). *Photographic Emulsion Chemistry*. Focal Press, London.

Kowaliski, P. (1972). *Applied Photographic Theory*. John Wiley and Sons, New York.

Kriss, M.A, and James, J.H., (ed.). (1977). *The Theory of Photographic Process*, 4th edn. Macmillan Publishing Co., New York.

Neblette, C.B. (1977). In: *Neblette's Hand book of Photography and Reprography*. Sturge, J.M. (ed.). Van Nostrand–Reinhold, New York.

Stroebel, L. (1990). *Basic Photography, Materials and Processes*. Focal Press, London.

12 PHOTOPOLYMER PLATES

INTRODUCTION

The production of printing plates for each of the major printing processes essentially involves the conversion of a photographic image on film into a relief, intaglio or lithographic image on metal or plastic plate. This is usually achieved by making use of the change in solubility of a light-sensitive coating on exposure to light. If this light-sensitive coating is exposed through a negative or positive, the areas receiving light usually become harder and less soluble. These areas may then be used as resist areas in etching the metal surface of the block, plate or cylinder, or else they may actually form the basis of a litho image.

The most important light-sensitive coatings used at the printing down stage of platemaking process include dichromated colloids, diazo compounds and photopolymers. Dichromated colloids consist of a dichromate salt combined with a natural or synthetic polymer (e.g. potassium dichromate and gelatin). Diazo compounds are the synthetic organic compounds used particularly in pre-sensitized lithoplates. Photopolymers may either be the materials in which polymerization of simple molecules (monomers) is influenced by the action of light or be the unsaturated long-chain polymers which on exposure to light cross-link to form larger molecules, with a lower solubility.

PHOTOPOLYMERIZATION

The polymerization reaction which is brought about by the influence of UV light is called photopolymerization and the polymer obtained is called photopolymer. This process requires the presence of a catalyst called photoinitiator which by the absorption of energy from UV light forms free radicals. These free radicals initiate the polymerization reaction.

The photopolymerization reaction involves two main mechanisms, namely, linear and cross-linking polymerization. In **linear polymerization**, a large number of simple molecules (monomers) under the influence of UV light undergo end-to-end linkage to form a long straight-chain polymer.

The **cross-linking polymerization** may occur along with linear polymerization. In this mechanism, one linear chain hooks on to another by bonding between chains resulting in a hard cross-linked polymer with network type of structure.

Photopolymerization initiated by visible light can also be performed in presence of a photosensitizer dye (such as xanthenes, thiazenes and acriflavine). This process is known as dye-sensitized photopolymerization which is gaining more importance in recent years. In photopolymer compositions employing sensitization, light is absorbed by the photosensitizer molecule thus causing an internal electron excitation. The electron energy is then transferred to the initiator which decomposes to form free radicals. Then, the free radical chemically attacks the monomer and initiates the polymerization reaction.

LITHOGRAPHIC PLATES

The conventional lithoplate is required to present two different types of surface to the printing system. The first is a hydrophilic surface which constitutes the non-image areas of the plate and is receptive to the aqueous dampening system. The hydrophilic surface of the plate is usually of grained and anodized aluminium. The oleophilic (ink-receptive) surface that constitutes the image areas of the plate is provided by the photosensitive coating. Thus in a lithographic plate the separation between the image and non-image areas is maintained chemically since they are essentially on the same plane. The wider the difference maintained between the ink receptivity of the image areas and the water receptivity of the non-image areas, the better the plate will be, the easier it will run on the press, and consequently, the better the printing. When a properly made plate is run on the press, the dampening rollers keep the non-image areas of the plate moist so that these areas do not accept ink. The ink roller then transfers the ink only to the dry image areas.

Most lithographic plates have a photosensitive coating that develops an image after being exposed through either a negative or a positive film, depending on the type of plate. Lithographic plates are conveniently grouped into two main types based on their way of exposure. The lithoplates that are exposed through photographic negatives are called negative working plates, whereas those exposed through photographic positives are called positive working plates.

Positive Working Plates

The compound commonly used for light-sensitive coating in a positive working plate is the diazo naphthaquinone (a derivative of naphthalene). When the coating is exposed to UV light in presence of trace amounts of water, nitrogen gas is liberated, and the original compound is converted into indene carboxylic acid.

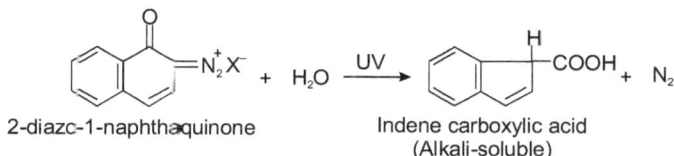

2-diazo-1-naphthaquinone   Indene carboxylic acid
(Alkali-soluble)

The indene carboxylic acid is soluble in an alkaline developer, while the unexposed coating remains insoluble.

During the platemaking process, the coating is exposed to UV light through a photographic positive, the image areas on the plate being protected from light by the silver image areas on the film. The non-image areas corresponding to clear areas on the film absorb UV light and the coating in these areas is converted to its alkali-soluble form. Finally, the plate is developed using an alkaline developer solution.

For long print runs, the finished plate is applied with a prebake solution and then baked in an oven at 230–250°C. The prebake solution contains synthetic detergents that prevent any oil or other materials from baking into the plate surface and sensitizing it. Baking polymerizes the coating in the image areas and makes it very hard and wear resistant. The baked coating is unaffected by the cleaning solvents used (such as blanket washes) and by the monomers used in the UV inks.

Negative Working Plates

The method producing an image on negative working plates is different from the method that is used on positive working plates. With negative working plates, the light makes the coating in the image areas hard and insoluble in the developer. This hardened coating remains on the plate as the ink-receptive areas.

Depending upon their action, the coatings applied to the negative working plates can be grouped into three types such as photoinsolubilization, photo cross-linking and photopolymerization coating.

Photoinsolubilization coatings   Diazo resin compounds are used as coating materials for negative working plates. These resins contain ionic diazo groups, which makes them water-soluble. Exposure to UV light brings about the removal of the ionic diazo groups from them and the coating becomes insoluble. The diazo resin is obtained by the condensation of diazonium salts such as 4-diazonium-1, 1´-diphenylamine chloride with formaldehyde. The resin so obtained is incorporated into the colloidal coating.

The photo cross-linking of the resin may also take place during exposure.

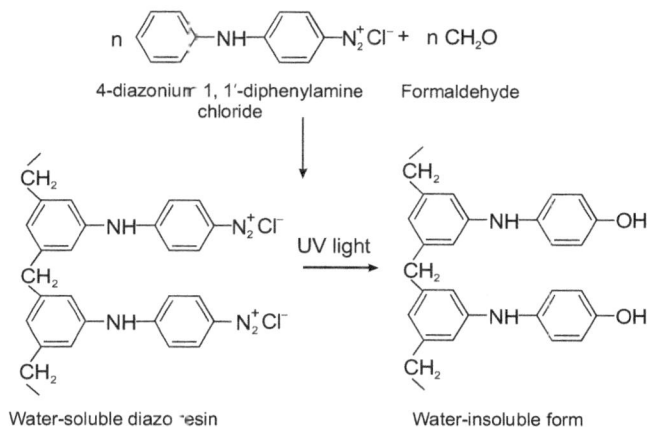

n 4-diazonium 1, 1'-diphenylamine chloride + n CH₂O → Formaldehyde

Water-soluble diazo resin   UV light →   Water-insoluble form

To increase the press run and photospeed of the coatings, polymeric binders were added. The binders have a higher light-sensitivity than diazo resin compounds and hence increase the photospeed of the coating. The diazo resin-coated negative working plates are used for medium run works such as newsprint and bookprint.

Photo cross-linking coatings   Certain organic compounds such as phenyl acrylic acid (cinnamic acid) dimerize when exposed to UV light.

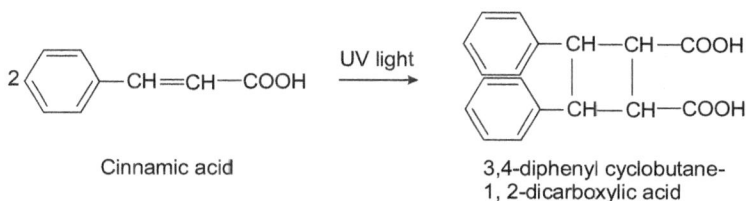

Cinnamic acid   UV light →   3,4-diphenyl cyclobutane-1, 2-dicarboxylic acid

This alkali and solvent-soluble phenyl acrylic acid and other acrylate derivatives (such as acrylamides) are incorporated into linear polymers. Photoinitiators are also added to the prepolymer. On exposure to UV light, the prepolymer chains cross-link through the acrylate groups resulting in a large increase in molecular mass. Due to this cross-linking, the coating material becomes hard and insoluble. Such highly cross-linked systems generally require solvent development, for example, polyvinyl cinnamate.

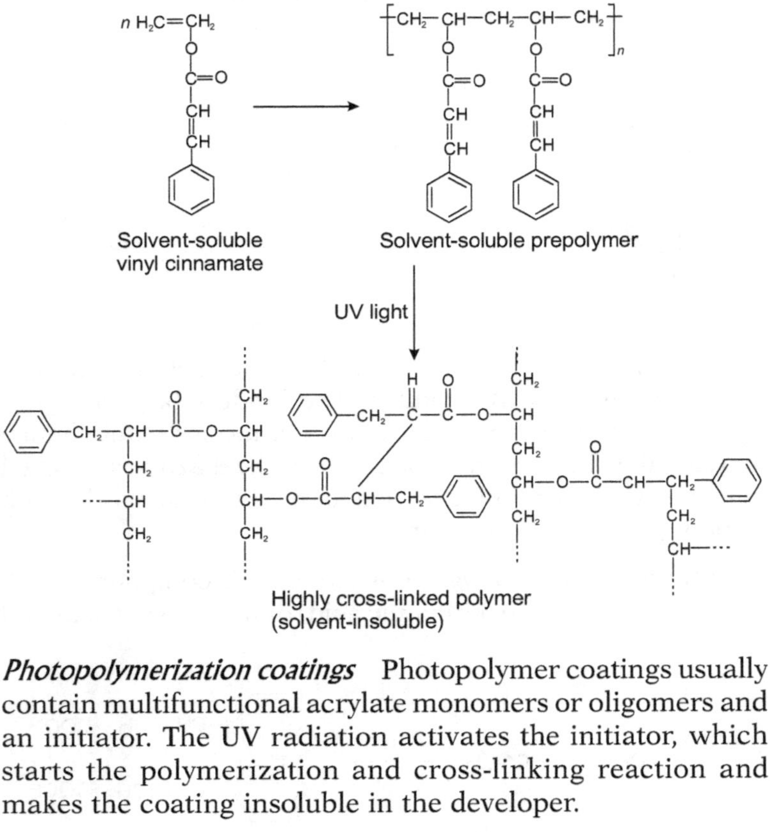

Solvent-soluble vinyl cinnamate

Solvent-soluble prepolymer

UV light

Highly cross-linked polymer (solvent-insoluble)

Photopolymerization coatings   Photopolymer coatings usually contain multifunctional acrylate monomers or oligomers and an initiator. The UV radiation activates the initiator, which starts the polymerization and cross-linking reaction and makes the coating insoluble in the developer.

Acrylamide in an aqueous alcohol-soluble form of nylon has been used to produce relief plates (Nyloprint). The exposed areas undergo polymerization and lose their alcohol-solubility.

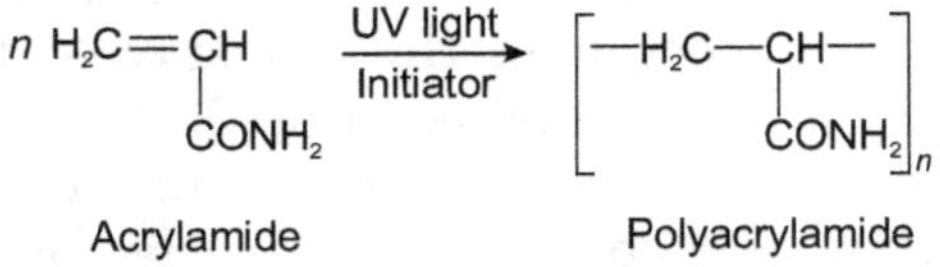

Acrylamide

Polyacrylamide

A multifunctional acrylate, pentaerythritol tetraacrylate is used along with the photoinitiator, chlorophenyltriazine.

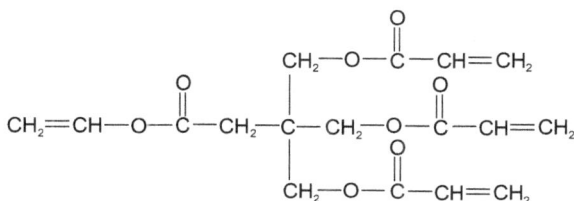

Pentaerythritol tetraacrylate

The photosensitivity of photopolymer system can be improved by the inclusion of an appropriate photo sensitizer such as Micheler's ketone.

Since photopolymerization is a photoinitiator-based radical reaction, it suffers from the disadvantage of susceptibility to oxygen quenching and peroxide termination mechanisms. In order to minimize oxygen inhibition, a barrier layer must be incorporated into the system. For this purpose, a coating of polyvinyl alcohol is provided at the surface.

Formulation of a photopolymer system	
Pentaerythritol tetraacrylate	Monomer
Urethane	Oligomer
Benzophenone	Initiator
Diphenyliodium hexafluorophosphate	Sensitizer
Polyvinyl butyral	Binder
Polyvinyl acetate	Oxygen barrier

FLEXOGRAPHIC PLATES

Flexography takes its name from the relatively soft and flexible plates used for printing. It is a form of rotary web letterpress using flexible plates (such as photopolymer or rubber) and fast drying or water-based inks. Originally, the flexography was used for paper bag printing but it subsequently proved ideally suited for the printing of almost all flexible packaging materials.

The development of photopolymer plates has promoted the growth of flexography, so that it is now being used not only in packaging, but also in publication printing. The photopolymer flexoplates are more suitable than rubber plates for jobs requiring close register, since the former is less flexible. They can be used with alcohol or water-based flexoinks. However, they are attacked by solvents such as ketones and aromatic hydrocarbons.

The advantages of photopolymer plates over rubber plates are the following.

- They are easily produced and uniform in thickness.
- They produce accurate, multicolour register and hold fine screens and small halftone dots.
- The production cost of a single plate is less than that of a rubber plate.
- The original used is a film which can be easily stored.

The disadvantages of photopolymer plates over rubber plates are the following.

- They conform less easily to the plate cylinder, since they are less flexible.
- They usually require a solvent wash to remove the unexposed photopolymer layer.
- The old plates are difficult to handle because they may curl, delaminate or tear when they are demounted or remounted from the plate cylinder.

Solid Photopolymer Plates

Solid flexo photopolymer plates are produced by cross-linking a rubber or elastomer with a photoreactive material in the presence of a photoinitiator. The structure of a typical solid photopolymer is shown in Figure 12.1. In this plate, a photopolymer layer is sandwiched between a protective polyester cover sheet or an alcohol-soluble protective sheet and a polyester backing that provides dimensional stability. A protective polyester cover sheet is provided at the top to exclude oxygen, because the radical-photopolymerization process is susceptible to termination through both oxygen quenching of the photoinitiator mechanism and by peroxide formation. An adhesive is used in some plates so as to promote the adhesion of the photopolymer layer to the backing.

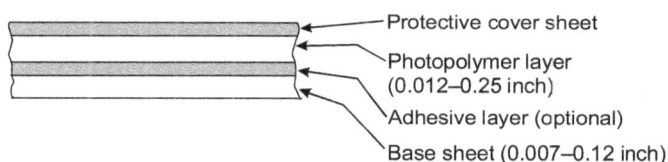

— Protective cover sheet

Photopolymer layer (0.012–0.25 inch)

Adhesive layer (optional)

Base sheet (0.007–0.12 inch)

Figure 12.1 Structure of a solid photopolymer plate

The photopolymer layer is usually made from a formulation that contains a thermoplastic elastomer, a photoreactive acrylate monomer, a photoinitiator and other additives. One popular photopolymer plate is the product of cross-linking of chloroprene (poly (2-chloro-1, 3-butadiene)) and trimethylolpropane triacrylate. On exposure to UV light, the trimethylolpropane triacrylate cross-links and hardens the polymer, making it insoluble. The degree of hardness introduced depends on the amount of triacrylate added to the formulation.

Another popular photopolymer formulation comprises styrene–isoprene rubber, trimethylolpropane triacrylate and a photoinitiator.

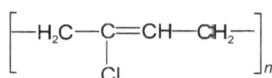

Chloroprene

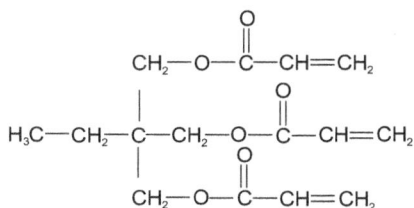

Trimethylolpropane triacrylate

314 Science and Technology of Printing Materials

Other acrylates used in the photopolymer formulation are ethylene glycol dimethacrylate, 1,4-butanediol diacrylate, 1,6-hexanediol dimethacrylate and triethylene glycol diacrylate.

A photoinitiator initiates and accelerates the polymerization of the monomer by the action of UV light. It absorbs energy from UV light and splits to form free radicals. These free radicals bring about the opening of the double bonds of the monomer, allowing polymerization to proceed. The commonly used photoinitiators are benzophenone, benzoin and benzyl.

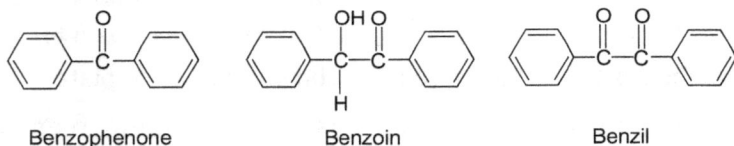

Benzophenone   Benzoin   Benzil

Under the influence of UV light, benzoin dissociates to form two free radicals:

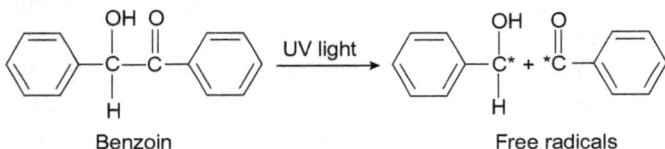

Benzoin   Free radicals

Plate exposure A two-step exposure is often used. The first exposure is made through the back of the plate without using a negative. This is to provide a solid base to the plate. The exposure is relatively short, usually about 1 minute.

The protective cover sheet is removed and the front surface of the plate is exposed to UV light through a photographic negative. The second exposure creates the relief image on the plate. The duration of the exposure determines the depth of relief produced but is usually of the order of 10 minutes.

UV light activates the photoinitiator, which starts the cross-linking reaction that turns the plate into a stable, insoluble material. The hardened exposed areas remain insoluble in the developer.

Processing The exposed plate is then developed by washing out the unpolymerized areas (unexposed areas) of the plate. The unpolymerized material is removed by gentle scrubbing with nylon brushes using a suitable solvent, usually an alcohol–ester solution.

The plate is then dried using a hot-air drier. This removes the residual solvents from the surface of the photopolymer layer and reduces the swelling to restore the image to their correct sizes. The final step is light finishing in which the plate surface is given a post exposure to UV light (250–260 nm). The main purpose of this finishing process is to remove the residual tackiness of the printing surface.

Liquid Photopolymer Plates

Liquid photopolymer plates are usually produced from partially polymerized acrylates or similar other chemically unsaturated products. The plate is provided with a polyester or stainless steel base. On exposure, the UV light hardens the coating material in the image areas without affecting the semi-liquid material in the non-image areas. The semi-liquid material in the non-image areas is then blown off the plate with a blast of air into a catch tank. The recovered material in the catch tank can be mixed with the fresh material and reused.

To obtain best results, the printing surface of the plate should be hard with a softer layer underneath. This can be achieved by providing at the top a thin layer of a resin composition that cross-links more extensively than the main layer beneath. This makes the printing surface of the plate hard with a softer layer underneath. The plates so obtained are known as **"capped plates"**. The hard surface layer reduces "squeezing out" and half-tone dots with holes in the centre and other typical problems in flexoprinting. The softer layer provides some cushion action to the plate, enabling it to print well on surfaces that are not very smooth.

REVIEW QUESTIONS

1. Explain the following terms.
   i. Photopolymerization
   ii. Photopolymer
   iii. Dye-sensitized photopolymerization
2. Discuss the mechanism involved in photopolymerization and dye-sensitized photopolymerization.
3. Describe the chemical nature of the light-sensitive coating on positive working lithoplates.
4. Explain the mechanism involved in the processing of positive working plates.
5. Mention the differences in the methods of production of positive and negative working lithoplates.
6. Explain how long-run positive working plates are produced.
7. With a neat diagram, explain the structure of a typical solid photopolymer plate.
8. Discuss the mechanism involved in the production of solid photopolymer plates.
9. Give the composition of photopolymer layer in a solid photopolymer plate.
10. Describe the method of exposure and processing of a solid photopolymer plate.

ADDITIONAL READING

Doyle, S. (1995). *Advances in Printing Plate Technology*. Pira International, UK.

METALS FOR PLATEMAKING

INTRODUCTION

Metals are the essential materials widely used for platemaking in printing industry. Their importance to the printer rests on the fact that in most cases printing ink is applied from a metal surface (e.g. copper or zinc plates, type alloys) and that in some cases printing ink is applied to a metal surface (e.g. aluminium foil, tin plate). This chapter deals with the properties of some important metals with reference to their uses in platemaking.

Criteria for the selection of metals for platemaking   In general, the metal suitable for platemaking is the one, which is malleable, with good wear resistance and strength and a suitably fine grain structure and is of reasonable cost. Other essential properties may include good thermal stability and the ability to be easily etched.

Metals like aluminium, copper, zinc and magnesium, which have proved themselves as the most suitable for platemaking. The thin layers of the metals like nickel and chromium are sometimes electroplated on to a printing plate to improve its wear resistance.

THE LITHOGRAPHIC PROPERTIES OF METALS

The basic requirement of a lithographic plate is that its image areas must be ink-receptive (i.e., oleophilic in nature) and non-image areas must be water-receptive (i.e., hydrophilic in nature). The relative wettability of metals in the lithographic process has been investigated by the measurement of contact angle. The contact angle is a measure of the liquid's ability to wet the surface.

CONTACT ANGLE

When a drop of liquid is placed gently on a slide, it either spreads out into a thin film or remains as a drop on the surface depending upon its affinity towards the solid surface. For this type of solid–liquid interface, interfacial tension can be expressed in terms of contact angle.

Contact angle, θ, is a quantitative measure of the wetting of a solid by a liquid. It is defined geometrically as the angle formed by a liquid at the three-phase boundary where a liquid, gas and solid intersect as shown in Figure 13.1.

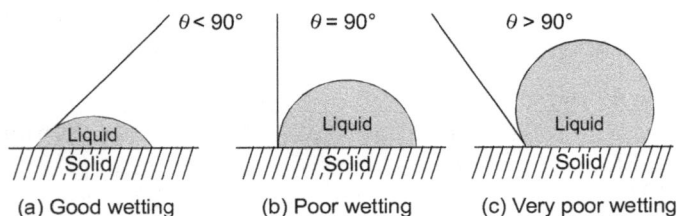

Figure 13.1 Contact angles

It can be seen from Figure 13.1 that low values of θ indicate that the liquid spreads or wets well, while high values indicate poor wetting. If the angle θ is less than 90°, the liquid is said to wet the solid. If it is greater than 90°, it is said to be non-wetting. A zero contact angle represents complete wetting.

On extremely hydrophilic surfaces, a water droplet will completely spread (an effective contact angle of 0°). This occurs for surfaces that have a large affinity for water (including materials that absorb water). On many hydrophilic surfaces, water droplets will exhibit contact angles of 10° to 30°. On highly hydrophobic surfaces, which are incompatible with water, one observes a large contact angle (70° to 90°). Some surfaces have water contact angles as high as 150° or even nearly 180°. On these surfaces, water droplets simply rest on the surface, without actually wetting to any significant extent. These surfaces are termed superhydrophobic and can be obtained on fluorinated surfaces (Teflon-like coatings) that have been appropriately micropatterned.

Measurement of Contact Angle

The contact angle of non-porous solids can be measured by the technique called goniometry. Analysis of the shape of a drop of test liquid placed on a solid is the basis for goniometry. The basic elements of a goniometer include a light source to illuminate the droplet, sample stage to hold the substrate, a syringe to apply a droplet of liquid, lens and image capture. Contact angle can be assessed directly by measuring the angle formed between the solid and the tangent to the drop surface.

After placing a droplet of liquid (1–25 μL) onto the surface, the outline of the droplet is studied through the magnifier. The operator positions the tangent line from the surface of the droplet at the point where it touches the solid surface. A protractor within the optics then provides a reading of the contact angle.

Unfortunately, operator subjectivity often interferes with the accuracy of the technique. Modern contact angle systems, however, adopt precision optics and charge-coupled device (CCD) cameras with image-processing hardware and software to enhance the performance of contact angle analysis, making it easier, quicker, and more precise.

A droplet of liquid is dispensed on to the substrate surface (manually or automatically), and a CCD camera reveals the profile of the droplet on the computer screen. Software calculates the tangent to the droplet shape and the contact angle. Data and the image are collected, analysed, and saved on computer.

Importance of Contact Angle

The contact angle values are useful in the selection of the best metals for image- and non-image areas of lithographic plates. For different metals, contact angles can be measured using water or other liquids such as oleic acid. The determination of contact angles of small drops of oleic acid applied to several metals that are previously immersed in water was carried out at Pira research lab in England. The results are shown in Table 13.1.

Table 13.1 Contact angles for different metals

Metal	Contact angle of oleic acid
Copper	18°
Aluminium	60°
Chromium	77°
Iron	178°

The low contact angle of oleic acid on copper indicates that copper is a good metal for the image areas of a lithoplate. On the other hand, the metals such as aluminium, chromium and iron are suitable for the non-image areas because the larger contact angles show that oil (i.e., ink) does not easily wet their surface.

(***Note*** Contact angle is usually measured on a smooth, well-cleaned metal surface under static conditions. However, the finished printing plates usually have a light-sensitive coating applied over the grained surface and they operate in press under dynamic rather than static conditions. So, the contact angles measured under ideal conditions are insufficient to predict the wetting quality of finished printing plates.)

METALS USED IN PLATEMAKING

COPPER (Cu)

Physical properties A reddish-brown metal which melts at 1084°C. It is malleable and ductile. It is a good conductor of heat and electricity. Its specific gravity is 8.95.

Chemical properties

Action of air It is not attacked by dry air at ordinary temperatures. But when heated to redness in air or oxygen, it first forms Cu_2O and finally CuO. In the presence of atmospheric carbon dioxide and moisture, it is covered with a green layer of basic carbonate, $CuCO_3 \cdot Cu(OH)_2$.

Action of acids Non-oxidizing acids (e.g. HCl and dilute H_2SO_4) have no effect on the metal in the absence of air or oxygen. In the presence of air, these acids dissolve the metal forming the corresponding salt and water.

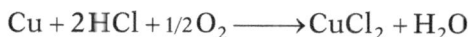

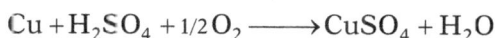

$$Cu + 2HCl + 1/2 O_2 \longrightarrow CuCl_2 + H_2O$$

$$Cu + H_2SO_4 + 1/2 O_2 \longrightarrow CuSO_4 + H_2O$$

Acids with oxidizing properties (e.g. HNO_3 and conc. H_2SO_4) readily attack copper.

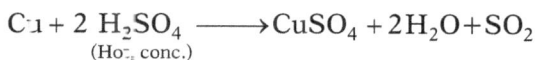

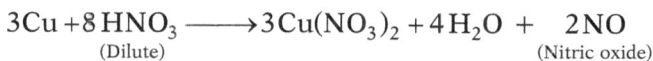

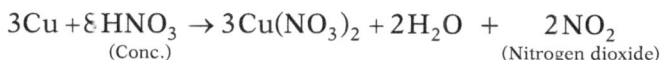

$$Cu + 2 H_2SO_4 \longrightarrow CuSO_4 + 2H_2O + SO_2$$
$$\text{(Hot, conc.)}$$

$$3Cu + 8HNO_3 \longrightarrow 3Cu(NO_3)_2 + 4H_2O + 2NO$$
$$\text{(Dilute)} \qquad\qquad\qquad\qquad \text{(Nitric oxide)}$$

$$3Cu + 8HNO_3 \rightarrow 3Cu(NO_3)_2 + 2H_2O + 2NO_2$$
$$\text{(Conc.)} \qquad\qquad\qquad\qquad \text{(Nitrogen dioxide)}$$

Action of water Copper is not attacked, even when hot, by water or steam.

Action of alkalis It resists the action of alkalis.

Uses

1. Copper is used in the production of printing surfaces for letterpress, lithography and gravure. It provides fine-line and half-tone blocks for the letterpress process, plates and cylinder for gravure.

2. The metal has a fine crystalline structure so that chemical etching or mechanical engraving can produce very small relief dots for letterpress or very fine recessed cells for gravure. Its oleophilic character makes it suitable to form the image areas of bimetal plates. The durability and affinity for ink of a deep-etch image can be improved by the application of a thin layer of copper and in the same way steel ink rollers can be made more inks-receptive.

3. It is also used in the production of electrotypes for duplicate letterpress plates.

4. The steel ink rollers can be made more ink-receptive by applying a thin coating of copper.

5. The durability and affinity for ink of a deep-etch image can be improved by the application of a thin layer of copper.

6. In the production of printed circuit boards and integrated circuits.

7. It is used for making a large number of alloys like brass, bronze, bell metal, etc.

ZINC (Zn)

Physical properties   Pure and polished zinc has a bluish white metallic lustre. It melts at 421°C. At ordinary temperatures it is brittle. On heating at about 155°C, it becomes malleable and ductile. Its specific gravity is 6.

Chemical properties

Action of air   On exposure to moist air, it slowly becomes tarnished with a thin film of the oxide. When heated strongly in the air, it burns with a bluish green flame and white clouds of ZnO are given off.

Action of acids Hydrochloric acid of all concentrations yields zinc chloride in solution, with hydrogen being given off.

$$Zn + 2HCl \longrightarrow ZnCl_2 + H_2$$

Sulphuric acid, when dilute, gives zinc sulphate and hydrogen gas, but with hot concentrated sulphuric acid, zinc sulphate and sulphur dioxide are formed.

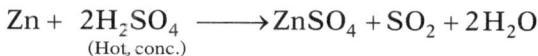

$$Zn + \underset{(Cold, dilute)}{H_2SO_4} \longrightarrow ZnSO_4 + H_2$$

$$Zn + \underset{(Hot, conc.)}{2H_2SO_4} \longrightarrow ZnSO_4 + SO_2 + 2H_2O$$

When zinc dissolves in nitric acid, zinc nitrate is formed in solution but the other products formed depend upon the dilution and temperature of the acid.

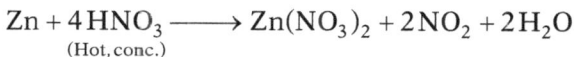

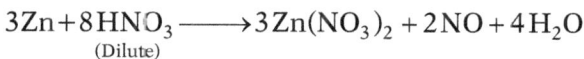

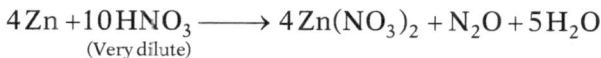

$$Zn + 4HNO_3 \longrightarrow \underset{(Hot, conc.)}{Zn(NO_3)_2} + 2NO_2 + 2H_2O$$

$$3Zn + 8HNO_3 \longrightarrow \underset{(Dilute)}{3Zn(NO_3)_2} + 2NO + 4H_2O$$

$$4Zn + 10HNO_3 \longrightarrow \underset{(Very\ dilute)}{4Zn(NO_3)_2} + N_2O + 5H_2O$$

Action of alkalis Zinc dissolves in hot concentrated caustic soda, liberating hydrogen gas and leaving sodium zincate in solution.

$$Zn + 2NaOH \longrightarrow Na_2ZnO_2 + H_2$$

Action of water Pure zinc has no action with water, but impure zinc decomposes steam quite readily with evolution of hydrogen gas.

$$Zn + \underset{(Steam)}{H_2O} \longrightarrow ZnO + H_2$$

Uses

1. Zinc is normally used to produce letterpress line blocks.

2. The corrosion resistance of iron or steel can be improved by the application of a protective coating of zinc, the process known as "galvanizing".

3. Used in the production of useful alloys such as brass, German silver, and delta metal.

4. Compounds of zinc such as ZnO, lithopone and zinc chromate are used as pigments.

ALUMINIUM (Al)

Physical properties   A bluish white metal, which is malleable, ductile, and easily machined and cast. It is a good conductor of heat and electricity. It melts at 660°C. It is a very light metal (sp. gr. 2.7).

Chemical properties

Action of air   Aluminium remains unaltered in dry air, but in moist air a film of oxide is formed at the surface, which protects it from further corrosion. It burns readily in air giving a brilliant light.

$$4Al + 3O_2 \longrightarrow 2Al_2O_3$$

Action of acids   The metal dissolves only slowly in dilute HCl, but quickly in the hot acid.

$$2Al + 6HCl \longrightarrow 2AlCl_3 + 3H_2$$

Sulphuric acid, only if hot and concentrated, attacks it, forming aluminium sulphate and sulphur dioxide.

$$2Al + 6H_2SO_4 \longrightarrow Al_2(SO_4)_3 + 3SO_2 + 6H_2O$$

Nitric acid of all concentrations has little or no effect on the metal. This is said to be due to the formation of a protective layer of aluminium oxide (i.e., the metal is rendered passive).

Action of alkalis   The metal is readily attacked by caustic alkalis, hydrogen being evolved and the corresponding aluminate left in solution.

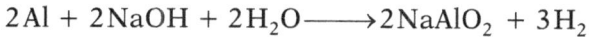

$$2Al + 2NaOH + 2H_2O \longrightarrow 2NaAlO_2 + 3H_2$$

Action of water   Pure aluminium is not affected by pure water. It decomposes boiling water liberating hydrogen.

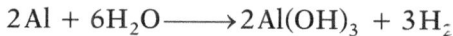

$$2Al + 6H_2O \longrightarrow 2Al(OH)_3 + 3H_2$$

Uses

1. Aluminium is widely used in making lithoplates.

2. The lightness and strength of aluminium alloys make them common components in a great deal of printing equipment including composing room furniture.

3. Aluminium foil is a very important packaging material.

4. Along with organic colloids, it is used for sizing of paper.

5. *Aluminium as a lithoplate material*   Aluminium is a good lithographic plate material for several reasons. It is reasonable in cost, light weight, available in uniform thickness and strong enough for the purpose. It is stiff enough to resist significant stretching when properly mounted on the press. It maintains good register of half-tone dots, even on large plates. One of the outstanding advantages of aluminium is its ability to protect itself against corrosion by the atmosphere. This is due to the formation of a protective oxide film that is self-healing when damaged. But the natural alumina film is extremely thin and does not confer wear resistance.

6. *Anodized aluminium*   Anodized aluminium provides an excellent surface for lithoplates with ceramic qualities that make it behave much like an old lithographic stone. Anodizing aluminium gives a thick alumina film (0.0025 to 0.030 mm). This protects the plate surface from mechanical damage and chemical attack, and greatly increases its press life. Thus lithoplates with improved

wear and corrosion resistance surface are produced from anodized aluminium. The water-holding properties of aluminium plates are achieved by a combination of increasing the surface area by graining and building a porous oxide film by anodizing.

An aluminium plate is anodized by connecting it as the anode of an electrolytic cell that commonly contains sulphuric acid as the electrolyte. Alternatively, phosphoric acid, chromic acid or oxalic acid can also be used as electrolyte. The coating forms as a result of progressive oxidation, starting at the aluminium surface. The thickness of the oxide film formed depends on the applied current density, temperature, and concentration of electrolyte. An oxide film, initially very thin, grows from the metal surface outwards and increases in thickness as oxidation continues at the anode. The outer surface of the oxide film is very porous and considerably softer than the layer beneath. The softer surface can be made harder and corrosion resistant by the process called "sealing". This process involves the treatment of the exposed film with boiling water. By this treatment, the porous alumina at the surface of the coating changes into its monohydrate form ($Al_2O_3 \cdot H_2O$), which occupies more volume, thereby the pores are sealed. Sealing can also be done by treatment of the oxide film with boiling dilute sodium dichromate solution.

Anodized film resists mild acids and bases, but it is attacked by strong acids and salt solutions. Anodized aluminium surface is not sufficiently water-receptive to fully reject the ink. The hydrophilic nature of the surface can be improved by treatment with an acidified solution of gum arabic or other hydrophilic polymers.

MAGNESIUM (Mg)

Physical properties   Magnesium is a light, silvery white and fairly tough metal. Its specific gravity is 1.74 and melting point is 650°C. It is fairly malleable and ductile.

Chemical properties

Action of air   Magnesium is stable in dry air but it is slowly tarnished in moist air due to the formation of the oxide, MgO. When it burns in air, it produces a bright white light.

$$2\,Mg + O_2 \longrightarrow 2\,MgO$$

Action of acids   Dilute acids, including nitric acid, attack magnesium to produce hydrogen.

$$Mg + 2\,HNO_3 \longrightarrow Mg(NO_3)_2 + H_2$$

Hot concentrated sulphuric acid produces sulphur dioxide.

$$Mg + 2\,H_2SO_4 \longrightarrow MgSO_4 + 2\,H_2O + SO_2$$

Action of water   Magnesium does not decompose water in cold condition. The reason is that a protective layer of magnesium hydroxide is formed on the surface of the metal which slows down the rate of further attack of water. However, it decomposes steam forming magnesium oxide and liberating hydrogen.

$$Mg + \underset{\text{(Steam)}}{H_2O} \longrightarrow MgO + H_2$$

Action of alkalis   Alkalis have no action on magnesium.

Uses

1.   Used in the graphic arts to make magnesium relief plates.
2.   Magnesium plates are used in photoengraving.
3.   Magnesium powder mixed with potassium chlorate is used in flash bulbs for photography. The flash powder burns quickly producing intense ultraviolet light which affects the photographic plate at once.
4.   Magnesium is used in making a number of useful alloys which are light, but strong and durable. These are used in several manufacturing applications, like high-volume parts including automotive and truck components.

5. Due to low weight, good mechanical and electrical properties, magnesium is widely used for manufacturing of mobile phones, laptop computers, cameras, and other electronic components.

CHROMIUM (Cr)

Physical properties   Chromium is a silvery white metal which melts at 1875°C. It takes a good polish which lasts long due to the formation of a self-protective oxide film. Its specific gravity is 7.2. It is hard and malleable.

Chemical properties

Action of air   It is not acted upon by air at ordinary temperature. When strongly heated in air, it forms chromic oxide.

$$4Cr + 3O_2 \longrightarrow 2Cr_2O_3$$

Action of acids   It dissolves slowly with dilute and concentrated hydrochloric acid, and with dilute sulphuric acid forming blue chromium(II) salts in solution.

$$Cr + 2HCl \longrightarrow CrCl_2 + H_2$$
$$Cr + H_2SO_4 \longrightarrow CrSO_4 + H_2$$

With hot and concentrated sulphuric acid, chromium(III) sulphate and sulphur dioxide are formed.

$$2Cr + 6H_2SO_4 \longrightarrow Cr_2(SO_4)_3 + 6H_2O + 3SO_2$$

No action with dilute nitric acid, but it becomes passive in concentrated nitric acid due to the formation of a protective chromic oxide film.

Action of alkalis   The metal is not attacked by alkalis under ordinary conditions.

Action of water   It decomposes steam at red hot, forming chromic oxide and hydrogen.

$$2Cr + 3H_2O \longrightarrow Cr_2O_3 + 3H_2$$

Uses

1. It is widely used in chromium plating because it gives a corrosion-resistant film.
2. Multi-metal plates use chromium for the non-image areas.
3. It is used in the preparation of chrome pigments (chrome yellow, chrome red, and molybdated chrome orange).
4. The most important use of chromium is in the production of special steels like ferrochrome and stainless steel.

NICKEL (Ni)

Physical properties   Nickel is a greyish white, lustrous and tenacious metal, capable of taking a high polish. It is hard, malleable and magnetic. It melts at 1453°C and its specific gravity is 8.9.

Chemical properties

Action of air   Nickel is fairly resistant to air at ordinary temperature. This is due to the formation of a thin protective film of oxide on the surface of the metal. When heated in air, it is slowly converted into the oxide.

Action of acids   Dilute hydrochloric acid or dilute sulphuric acid slowly dissolves the metal.

Dilute nitric acid dissolves the metal rapidly. However, the metal becomes passive in the presence of concentrated nitric acid.

Action of water   Red hot nickel decomposes steam.

$$\underset{\text{(Steam)}}{Ni + H_2O} \longrightarrow NiO + H_2$$

Action of alkalis   Nickel is resistant to the action of alkalis.

Uses

1. Used in nickel-plating, which provides a hard, corrosion and wear-resistant coating.

2. Nickel, in a finely divided state, is used as a hydrogenating catalyst for hardening of oils.

3. It is widely used in making alloys such as nickel steel, German silver, nichrome, etc.

LEAD (Pb)

Physical properties Lead is a soft bluish grey metal, which melts at 327.5°C. It is malleable but not very ductile. Its specific gravity is 11.34.

Chemical properties

Action of air It is not attacked by dry air but gets tarnished in moist air due to the formation of a thin film of the basic carbonate. When the metal is heated in air, oxides of lead (PbO, Pb_3O_4) are formed.

Action of acids Lead is only slightly attacked by dilute hydrochloric acid and sulphuric acid at ordinary temperature. This is due to the formation of an insoluble film of lead chloride or lead sulphate on the surface which resists further action. Hot concentrated sulphuric acid readily dissolves the metal with the evolution of sulphur dioxide.

$$Pb + 2H_2SO_4 \longrightarrow PbSO_4 + SO_2 + 2H_2O$$

When lead dissolves in nitric acid, lead nitrate is formed in solution but the other products formed depend upon dilution of the acid.

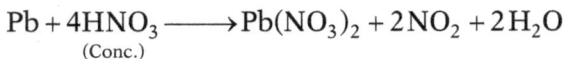

$$\underset{\text{(Conc.)}}{Pb + 4HNO_3} \longrightarrow Pb(NO_3)_2 + 2NO_2 + 2H_2O$$

Action of alkalis It is slowly attacked by caustic alkalis with the evolution of hydrogen.

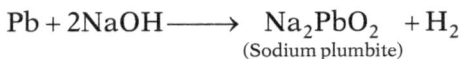

$$Pb + 2NaOH \longrightarrow \underset{\text{(Sodium plumbite)}}{Na_2PbO_2} + H_2$$

Action of water Pure air-free water has no action on the metal. It slowly decomposes steam at high temperatures.

It is readily corroded by water containing dissolved air forming lead hydroxide in solution.

Uses

1. Used in the preparation of pigments like red lead, white lead, chrome red, chrome yellow, etc.
2. Used in making useful alloys such as type metal (an alloy of lead (75%) and antimony (20%), which is used for making printing type), and solder (used in soldering).
3. A major constituent of the lead-acid battery used extensively in car batteries.
4. Used as shielding from radiation.

GRAINING OF LITHOPLATES

Graining is a method of roughening of the surface of metal plates. This process increases the surface area and the increase is between two or four times the area of a smooth plate.

PURPOSE OF GRAINING

There are many reasons for graining lithoplates.

- Graining improves the adhesion of image coatings and other films, giving them longer life.
- Grained plates have more latitude on press and hold water better on the press. Thus dampening becomes easy with grained plates.
- Grained plates are more durable.
- The possibility of ink emulsification or plate dry up is decreased by graining of lithoplates.

METHODS OF GRAINING

Graining can be carried out mechanically, electrochemically, chemically or by combinations of these.

Mechanical Graining

The brush graining is the commonly used mechanical graining method. First the plates are subjected to the pre-cleaning process and then passed through a brush-graining machine. The brushes are wet with a continuously recycled abrasive, often a mixture of pumice and sand. The rotating, oscillating brushes support the abrasive particles as they impact the surface, giving a pattern of fine scratches in the surface. Three or four brushes produce a better, more uniform surface, especially at high web speeds. As the web continues through the machine, it is often lightly etched with a solution of strong alkali to remove embedded grit, and carefully rinsed with distilled water.

In general, brush graining is smoother and inferior to electrochemical graining but is still widely used because it is fast and relatively inexpensive. Brush grains are often improved by chemically graining the brush-grained surface.

Electrochemical Graining

It is carried out in an electrolytic cell using dilute hydrochloric acid or nitric acid as the electrolytic solution. An alternating current is applied between the electrodes. The operating temperature is maintained between 20 and 40°C. Graining is generally performed by imposing an alternate current on a moving web, using stationary electrodes and very high current density for short process time. Electrochemical graining is more versatile as it produces a rougher and more uniform surface. However, the method is more expensive.

Chemical Graining

In this process, graining is carried out by the action of suitable chemicals. The chemical used dissolves the surface oxide film and allows the raw metal to chemically react to produce the grain. Lithoplates can be chemically grained using trisodium phosphate (Na_3PO_4) or ammonium bifluoride (NH_4HF_2).

However, chemical graining method is not found to be successful.

REVIEW QUESTIONS

1. What are the criteria for the selection of metals for platemaking?
2. Define the term "contact angle". Describe the determination of contact angle of a liquid.
3. Discuss the technological importance of contact angle.
4. How and under what conditions does copper react with
   i. Air
   ii. Nitric acid
5. Give the applications of the metal copper in the printing industry.
6. Describe the reactions of zinc with:
   i. Sodium hydroxide
   ii. Nitric acid
   iii. Sulphuric acid
7. What happens when aluminium foils are
   i. treated with dilute hydrochloric acid
   ii. placed in a solution of copper sulphate.
8. Compare the chemical properties of aluminium and copper.
9. Write a note on "Anodized aluminium".
10. Give the reaction of lead with
    i. Nitric acid
    ii. Silver nitrate solution
11. What happens when magnesium foils are treated with
    i. Hot concentrated sulphuric acid
    ii. Nitric acid

12. Give the applications of the following metals in the printing industry:
    i. Aluminium
    ii. Zinc
    iii. Magnesium
    iv. Chromium
13. Discuss the physical and chemical properties of the metal chromium.
14. Explain the action of lead with
    i. Concentrated nitric acid
    ii. Hot concentrated sulphuric acid
    iii. Sodium hydroxide solution
15. What are the main objectives of graining of lithoplates?
16. Describe the method of mechanical graining of lithoplates.
17. How is electrochemical graining of lithoplates performed? Mention the advantages of electrochemical graining.

ADDITIONAL READING

Berg, J.C. (1993). *Wettability*. Marcel Dekker, New York.

Cotton, F.A. and Wilkinson, G. (2004). *Advanced Inorganic Chemistry*, 5th edn. John Wiley and Sons Ltd., USA.

Lee, J.D. (2003). *Concise Inorganic Chemistry*, 5th edn. Blackwell Science Ltd., London.

Schrader, M.E. and Loeb, G. (1992). *Modern Approach to Wettability*. Plenum Press, New York.

Shriver, D.F. and Atkins, P.W. (1999). *Inorganic Chemistry*, 3rd edn. Oxford University Press.

14   PRESS ROLLERS

INTRODUCTION

Most of the printing processes except screen-printing use rollers in some way or the other. In offset, letterpress, and collotype printing, rollers are used to transport the ink from the ink fountain to the form rollers. The transfer of ink from the fountain to the surface to be printed is an important factor in controlling both the quality and production in the printing process. A simple inking system with one or two rubber rollers is used in flexographic process. In the conventional and electro-assist gravure process, a special rubber impression roller is used. Rollers are even used to transport the paper and distribute ink in electrostatic printing.

GENERAL REQUIREMENTS OF ROLLERS

Unfortunately, there is no one single type of roller that will fulfil all the requirements needed in the various printing processes. However, in general, rollers should have the following characteristics to provide a better performance.

- Good affinity for ink and better ink transfer in the case of inking rollers and a good affinity for water in the case of dampening rollers.
- Good affinity for water
- Good solvent resistance
- Good heat resistance even at high working speed
- Good abrasion resistance

- Good resilience
- Good tensile strength, hardness and dimensional stability
- Good wear resistance properties
- Should be easily washable

The proper selection of roller materials is essential for better performance. In such selections, consideration must be given to the hardness desired (durometer), position and functions of the roller on press, press speed, the type of ink and the solvents which will come in contact with the roller, and the printing method.

TYPES OF ROLLERS

Rollers can be classified based on the chemical nature of the materials used in their production.

Composition Rollers

The composition rollers show better performance at lower speed and lower temperature conditions. They distribute the ink well and hold the lint dust, thus eliminating specks and spots when printing solids. These rollers exhibit lower tensile strength and are sensitive to high temperature. Composition rollers cannot be used in high-speed presses because they readily disintegrate due to the heat build-up and lack of tensile strength. Earlier, glue glycerin composition rollers were used in letterpress. Now composition rollers are rarely used in the printing industry.

Synthetic Rubber Rollers

They show satisfactory performance with almost all methods of printing and with most types of printing inks. Synthetic rubbers (or elastomers) such as Buna-N, butyl rubber and neoprene rubber are the commonly used roller materials. It is important, however, to match the rubber polymer with ink, solvents and presses to ensure long life. This type of roller is extensively used on very high-speed presses. The effect of internal friction observed in these is lesser than that in plastic and PVC blend rollers.

Plastic Rollers

The main plastic materials that are widely used in making press rollers are polyvinyl chloride and polyurethane. These materials also exhibit the good features of rubber with characteristics unique to plastic. They have generally replaced composition rollers in sheet-fed letterpress printing by providing the good features of composition materials added with high abrasion resistance. They have also been used as substitutes for rubber in sheet-fed lithography because of their high tensile strength, good abrasion and solvent resistance.

Rubber/Plastic Blend Rollers

They are mostly used in lithographic and letterpress rollers. PVC (polyvinyl chloride) is often used as the plastic component in the blend. Such rollers provide good affinity for ink, tensile strength, dimensional stability and wear resistance.

Steel Rollers

Steel rollers are used in letterpress and offset printing. Basically steel is more water-receptive than ink-receptive, hence it accepts ink if ink is applied first. Thus, in lithographic printing, steel rollers accept ink only if the ink is applied before any desensitizing material such as gum arabic comes in contact with steel. Under certain conditions, the gum in the fountain solution may get adsorbed on to the steel in some areas and this may result in stripping of rollers. The steel does not accept ink in these areas.

The conditions such as excess gum or acid in the fountain solution and the use of excess fountain solution may result in stripping of steel rollers. The ink-stripping problem of steel rollers can be avoided in two ways. One way is to use a roller that is covered with an ink-receptive layer made up of a highly cross-linked hard rubber or a plastic material such as nylon. The other way is to cover the steel with a layer of copper, which is more ink-receptive than steel. A thin layer of copper can be chemically deposited by treating the cleaned roller surface with a solution containing cuprous chloride (Cu_2Cl_2) and hydrochloric acid dissolved in a mixture of ethylene glycol and isopropyl alcohol. The cuprous ions are reduced to metallic copper by the iron atoms of the steel as per the reaction

$$2Cu^+ + Fe \longrightarrow 2Cu + Fe^{2+}$$

A very thin layer of copper is formed, which may wear off easily and hence the treatment must be repeated frequently.

Ceramic Rollers

Ceramic rollers are used as water rollers (dampening rollers) in offset printing. These smooth and glossy rollers are hydrophilic (water-loving) in nature. They usually carry a very thin film of water and hence resist ink. They can be suitably used in place of chrome dampening rollers. They help to reduce or eliminate the streaking and are useful in running alcohol substitutes. They resist scoring and grooving, because of their extreme hardness. Ceramic rollers usually exhibit better wear resistance. Ceramic anilox rollers in flexography resist wear. Otherwise wear may give rise to serious problems on repeat runs because it changes the amount of ink carried and also changes the colour of the print.

ANATOMY OF A ROLLER

If a new roller is observed under a microscope, it looks very uneven, full of fingers, ridges and gorges, almost like sawtooth mountain ranges. The roughness is created when the roller is made, and the rubber stretches and breaks around the grinding wheel. The "mountains" on the surface greatly increase the surface area of the roller. This rough surface carries ink and water, and the constant agitating motion of the fingers keeps the ink and water in a uniform emulsion. The surface of a roller is continuous with ridges and valleys, and is entirely flexible.

Nap

The rough surface of a good roller is normally called the nap. A good nap is essential for good printing. One can test for sufficient nap by pushing the finger away down the axis of the roller. If the roller catches the finger, it has nap and will print well. Some rubber materials have little nap. If the roller is too smooth, it is said to be a defective roller for offset.

Glaze

With continuous use, rollers tend to get glazed, which means the mountains become filled and sealed with particles from the materials such as inks and fountain solutions that come in contact with the roller (Figure 14.1). Thus, glaze is caused by the build-up of particles in the nap of the roller surface. This gives the roller a smooth surface and makes it unable to convey ink or water, which results in a long list of printing problems.

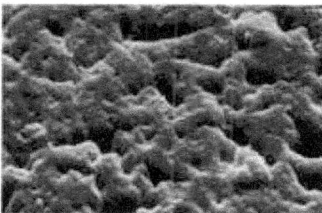

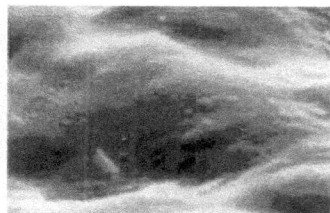

Clean roller   Glazed roller

Figure 14.1 Anatomy of clean and glazed rollers

When rollers are glazed, the "fingers" in the rubber surface do not transfer ink effectively, which means the pressmen will need to run more ink and more water on the press to achieve the required ink densities. Excess of ink and water will result in dot gain, mottled solids, hickeys, ink-drying problems, reduced quality, and wastage of ink and paper. A glazed roller requires proper cleaning for better performance. One can also use Bottcher's glaze-free rollers which run much longer than conventional rollers without developing glaze, and they are easier to clean.

CLEANING OF ROLLERS

After a long run, press rollers require proper cleaning so as to remove the undesired deposits, such as dried ink and glaze from the surface. This is essentially required for better performance of the roller and to get quality printing. The cleaning of rollers is usually carried out by using suitable solvents or solutions. Solvents that are powerful in dissolving dried oil-based inks also dissolve or soften rubber rollers. Therefore, a press cleaner solution requires proper formulation and compounding so as to remove only the undesired deposits from the surface without causing any softening or swelling of the roller material. The solvents used must not evaporate too rapidly or too slowly and must be less toxic in nature.

Press cleaners can be conveniently divided into two types: one is to remove wet ink and the other to remove glaze.

Removing wet ink   Naphtha is a good wet ink cleaner. It is less toxic and has a flash point of 11–12°C. It shows a least swelling effect on rollers. The mineral spirits (flash point 38–48°C) except kerosene and gasoline can also be used. Special types of two-solution cleaners are also used for the efficient cleaning of ink rollers. The advantage of two-solution cleaners is that they not only remove the wet ink but also remove ink of other colours that were run previously.

The first solution used is an aqueous emulsion containing a detergent and an aliphatic hydrocarbon solvent. After the removal of most of the ink by the first solution, the second solution is applied. Hydrocarbon solvents such as naphtha or mineral spirits are used. However, it is important that sufficient amount of the final solution is to be used to remove the previous solution completely. Otherwise, the rollers will not take up the ink properly.

Removing glaze   The nap of a glazed roller is filled and covered with different particles accumulated from all of the materials that pass over the roller. This includes water-soluble particles (like gums, coatings, dust), solvent-soluble particles (like ink, etc.), and insoluble residues (like mineral deposits, clays, soap deposits). Calcium from the paper is particularly annoying when it collects on rollers. Thus a uniform white glaze develops on the surface of rollers in either inking or dampening trains. This gives the roller a smooth shiny surface and interferes with the ability of the rollers to carry uniform ink and/or water films. This leads to poor ink transfer, roller stripping or streaking. In extreme cases, a hard band of white calcium compounds builds up at roller ends, and roller settings cannot be properly maintained. The regular cleaning also eliminates calcium compounds.

A three-step cleaning system is usually employed for the proper cleaning of rollers. The steps involved are the following.

Step 1   Initial washing is carried out with a water-miscible wash solution (solvent mixed with water). This removes the top layer of solvent-soluble and water-soluble particles from the nap of the roller.

Step 2   A non-grit roller paste is then applied to remove the insoluble particles.

Step 3   Final washing is carried out once again with a water-miscible wash to remove the remaining soluble particles.

APPLICATIONS OF ROLLERS IN PRINTING

The selection of suitable rollers for any printing job mainly depends on the method of printing employed and the type of ink used. Press rollers are mainly used as ink rollers and dampening rollers. Ink rollers have to perform two functions in the transfer of ink to the paper. First one is to deposit a uniform layer of ink from the fountain to the plate and the second one is to continue the agitation process whereby the ink is presented to the plate in proper consistency. This is generally accomplished by using a series of elastomeric and steel rollers. Duck roller picks up ink from the fountain and passes it on to the distributor rollers which perform a continuation of the agitation process. Form rollers accept the ink from the distributor rollers and present it to the plate in a thin uniform film.

In Offset and Letterpress

The rollers used for sheet-fed offset, web offset and publication printing are usually made up of synthetic rubbers (such as Buna-N) or rubber/plastic blends (such as nitrile/PVC blend). These rollers exhibit excellent surface tensile strength to withstand the vehicles used in most common inks except UV inks. Urethane materials with improved dimensional stability are useful as cover rollers. They are tough and are not affected by the solvents that are used in offset and letterpress inks. Different types of rollers are used in dampening system on lithographic presses. Most lithographic presses use separate distribution systems with cloth-covered rollers to transfer the fountain solution to the plate. In the "bareback" system rubber or urethane rollers are used. Neoprene and urethane rollers are usually used for letterpress news inks. Urethane resists swelling, and neoprene is flame retardant so that the roller will not contribute to potential flash or fire. Nitrile/PVC rollers are widely used on presses running web offset news inks. Buna-N, nitrile/PVC and urethane rollers are useful on moisture-set, water-washable and oil type letterpress inks for printing Kraft substrates.

In Flexographic Presses

Many different types of inks are used in flexography since the process is employed to print on a wide variety of substrates. Therefore, the flexographic roller materials must be compatible with the ink being used. For example, natural rubber or neoprene rollers are suitable for water or alcohol-based inks Ethylene/propylene synthetic rubber rollers are used for inks with high concentration of acetates or ketones. Rollers made with Buna-N are used for inks containing aliphatic hydrocarbon solvents. Flexographic rollers must have finely ground polished surfaces. They must be blemish-free and must have good rebound characteristics.

In Gravure Process

The basic covering materials used on gravure are Buna-N and neoprene with durometers of about 75–95, depending on the texture of the substrate being printed. The use of electrostatic-assist process in gravure printing requires rollers with good electroconductive properties. They are typically known as semi-conductive impression rollers. This roller is covered with a rubber coating that is 0.5–1.0 inch thick. It is made semi-conductive by the inclusion of special carbon blacks, or a long-chain amino or hydroxyl polymer. Most of the gravure presses are now equipped with an electrostatic assist.

In Ultraviolet Inks

Printing with UV-curing inks requires the use of specially formulated rollers. The vehicles and the photoinitiators used in UV inks usually cause the swelling of the conventional rollers. The synthetic rubbers used in conventional rollers are affected by the polar vehicles present in UV inks. Blends of synthetic elastomers are the most suitable roller materials for use with UV inks. In printing, where both UV-curing and regular inks are to be used on the same press, rollers would normally be of the nitrile/PVC type. Urethane rollers are generally not suitable for printing with UV inks. The specially formulated roller materials such as ethylene/propylene synthetic rubber are more effective for use with UV inks.

In Removing Hickeys

Sometimes the matter being printed may develop defects like **hickeys**. These are small areas that print almost solid in the centre and are surrounded by a white ring or halo. They are caused by small pieces of ink skin, or other particles that become attached to the plate, or on lithographic press, to the blanket. They can also be caused by a rubber roller that fails to start. They print dark because ink skin and rubber are ink-receptive. The ink cannot print around the edges of the thick pieces and this situation creates the white ring. The coating particles may occasionally accept ink and also produce hickeys.

Hickey rollers are specially designed to remove hickeys from lithographic printing plates. Different types of hickey removal systems are employed in modern presses. One of the popular types has a nap or suede finish that brushes or wipes debris off the plate.

REVIEW QUESTIONS

1.   Discuss the general requirements of a roller.
2.   Describe the special features of
   i.   Composition rollers
   ii. Synthetic rollers
   iii. Plastic rollers
   iv. Steel rollers
   v.   Ceramic rollers

3. How can the stripping problem in steel rollers be avoided?

4. What are the specifications for the roller materials to be used in the following cases?

    i. Flexographic process

    ii. Gravure process

    iii. For UV inks

    iv. Offset and letterpress

5. Write a note on rollers used for hickey removal.

6. Discuss the anatomy of a roller.

7. Explain the terms:

    i. Nap

    ii. Roller glaze

8. Describe the method of cleaning of press rollers.

ADDITIONAL READING

NIIR Board. (2002). *Hand Book on Printing Technology (Offset, Gravure, Flexo, Screen)*. National Institute of Industrial Research, New Delhi.

NIIR Board. (2006). *A Complete Book on Printing Technology*. Asia Pacific Business Press Inc., New Delhi.

15   FOUNTAIN (DAMPENING) SOLUTIONS

INTRODUCTION

In lithographic printing, it is essential to keep the non-image areas of a plate moistened with water so that they will not accept the ink. During the process of platemaking, the non-image areas are desensitized by treating with an acidified solution of gum arabic. A thin adsorbed film formed by the gum arabic desensitizes the non-image areas. If the desensitized film were able to remain on a plate indefinitely, it would be possible to run the plate with nothing but water in the dampening fountain. When plates are running clean, the gum film may gradually wear off, but it is replaced with the gum from the dampening solution.

The dampening of the offset plate is one of the most critical factors in the production of quality offset printing. Through the use of a fountain concentrate mixed with water, a working fountain solution is produced that performs several essential functions on press.

COMPOSITION OF FOUNTAIN SOLUTIONS

The following are the main ingredients used in fountain solutions.

Desensitizing Gum

The function of gums is to adhere to the plate's non-image area and protect it from accepting the ink. Water-soluble and film-forming gums are widely used for this purpose.

Chemically a desensitizing material must be a hydrophilic or a water-receptive material. Desensitizing materials do not react chemically with the plate metal but are adsorbed on the surface. The gum also serves to protect the plate from humidity and chemical attack possible during its shelf life. Gum arabic is the commonly used desensitizing material. Other water-soluble polymers such as larch gum, oxidized starches, carboxy methyl cellulose (CMC), polyvinyl pyrrolidone, and polyvinyl alcohol have also been found useful as plate desensitizers.

Acid

It is mainly added to reduce the pH of a fountain solution. The acid in the fountain solution ensures that the action of gum is reinforced and not destroyed. Thus it helps in keeping the plate image-area sensitive to ink and non-image area sensitive to water. The commonly used acids in fountain solutions are phosphoric acid, citric acid and lactic acid. Gum arabic functions well as a desensitizing gum in phosphoric acid medium. Addition of phosphoric acid to a solution of gum arabic in water converts the calcium, potassium and magnesium salts of arabic acid into free arabic acid.

$$R - COOK + H^+ \longrightarrow R - COOH + K^+$$
$$\text{(Gum arabic)} \qquad\qquad \text{(Arabic acid)}$$

The highly polar carboxyl (-COOH) groups of arabic acid are responsible for its adsorption on to the metal plate. This results in the desired hydrophilic surface of the non-image areas. In the right range, acids act as mild detergents and prevent the build-up of the ink on the plates. Too much acid, however, can cause ink sensitivity because it will over-etch the plates.

Wetting Agents or Surfactants

They lower the surface tension of water allowing it to maintain the wetting characteristics of the non-image areas of the plate. By reducing the amount of water required to keep the plate clean, they also reduce the amount of ink required for printing. Isopropyl alcohol is the widely used wetting agent.

Water

It is the main constituent (about 95–98%) of a fountain solution.

Plate Conditioners/Additives

The fountain solution also contains other additives that are added to meet certain specific requirements.

Fountain solution is usually added with a buffering agent so as to keep the pH stable during the course of printing. Ideal pH for most acid fountain solutions lies in the range of 4.0–5.0. Some additives are added to minimize the corrosive action of the acid on the metal plate. This will extend the plate life and improve the overall printing quality. Magnesium nitrate [$Mg(NO_3)_2$] serves partly as a buffer and partly to reduce plate corrosion. Non-piling agents are added to eliminate any chances of piling. Typical non-piling agents are made from glycols that will tend to keep the blanket moist. Silicone materials are added to enhance the release characteristics of the blankets and minimize piling due to paper picking tendencies, etc. Fungicides are added to keep the fountain solution free from mildew.

In general, the gum, wetting agent, acid and additives are combined to form what is known as "fountain concentrate" or "fountain etch". This solution is mixed with water at the press side or at the central area and then used.

Isopropyl Alcohol (IPA)

Isopropyl alcohol, the additive of printers' choice in certain dampening systems, is a target of increasingly stringent environmental regulations because it is a volatile organic compound. In the early 1990s, most printers were using alcohol substitutes to reduce or replace IPA on both web- and sheet-fed presses. Alcohol is not needed in conventional dampening systems as they use the paper sleeves or fabric covers. Alcohols or alcohol substitutes work best in continuous dampening systems, which have roll-to-roll contact.

Alcohol substitutes are all proprietary mixture of solvents. Even though alcohol substitutes are proprietary mixtures, they do not have a couple of common traits that should be considered: First they are strong ink solvents and secondly, they are very non-volatile. Because of this combination of properties, they tend to remain in the fountain solution and, at the same time, can extract some oil portions from the ink. Eventually, this can contaminate the dampening system. For this reason, it is very important that the dampening system should be thoroughly cleaned on a regular basis.

Some of the commonly used alcohol substitutes are summarized in Table 15.1.

Table 15.1 Commonly used alcohol substitutes

Chemical name	Formula
Ethylene glycol	$HO \cdot CH_2 \cdot CH_2 \cdot OH$
Butyl ether of ethylene glycol	$C_4H_9 \cdot O \cdot CH_2 \cdot CH_2 \cdot OH$
Diethylene glycol	$HO \cdot CH_2 \cdot CH_2 \cdot O \cdot CH_2 \cdot CH_2 \cdot OH$
Diethylene glycol ethyl ether	$CH_3 \cdot CH_2 \cdot O \cdot CH_2 \cdot CH_2 \cdot O \cdot CH_2 \cdot CH_2 \cdot OH$
Triethylene glycol	$HO \cdot CH_2 \cdot CH_2 \cdot O \cdot CH_2 \cdot CH_2 \cdot O \cdot CH_2 \cdot CH_2 \cdot OH$
Propylene glycol	$CH_3 \cdot CH(OH) \cdot CH_2 \cdot OH$
2-Ethyl-1, 3-hexanediol	$HO \cdot CH_2 \cdot CH_2 \cdot CH_2 \cdot CH(OH) \cdot CH(C_2H_5)CH_2 \cdot OH$

As an ingredient of fountain solutions, isopropyl alcohol (IPA) performs a variety of different functions that are discussed below.

- It reduces the surface tension of water, allowing water to wet the dampening roller more evenly. As a result, the amount of fountains required is reduced and water spreads more evenly to the plate from the rollers.

- It increases the viscosity to provide a thicker layer of fountain solution to be applied across the rollers, thereby improving the performance of the ink, paper and printing plates.

- It reduces the emulsification of ink into water and water into ink.

- It maintains better ink–water balance.

- It allows for less moisture to be carried to the paper, thereby causing less ink drying problems.

- It acts as a lubricant, thereby reducing the drag between the form rollers, inking rollers and the plate.

Some of the disadvantages of using IPA in a fountain solution are the following.

- It is a volatile organic compound that causes air pollution when released to the atmosphere. Hence, its use may trigger regulatory obligations.

- It has low flash point (21°C) and hence must be handled with extreme caution.

- Its fume can cause irritation, without proper ventilation in the pressroom.

- It is usually more expensive than most alcohol substitutes.

TYPES OF FOUNTAIN SOLUTIONS

Based on the chemical nature (mainly pH), fountain solutions may be grouped as acid, alkaline and neutral fountain solutions.

Acid Fountain Solutions

These fountain solutions usually contain an acid such as phosphoric acid, citric acid or lactic acid. Phosphoric acid is commonly used in most of the acid fountain solutions.

These acid fountain solutions are usually employed in lithographic printing process. The acid converts gum arabic (the desensitizing gum) to its free acid form (arabic acid) in which the molecules contain carboxyl groups. These groups help the gum to adsorb to the plate surface. Phosphoric acid not only acts as an acid, but also shows the desensitizing properties. Sufficient amount of acid should be used to convert most of the gum into its free acid form. But excess acid can lead to the corrosion of plate and actually can cause sharpening of dots on a lithographic plate.

Alkaline Fountain Solutions

Alkaline fountain solutions are particularly useful in offset newspaper printing. They are made up of phosphates such as disodium hydrogen phosphate (Na_2HPO_4) and sodium dihydrogen phosphate (NaH_2PO_4). Such solutions do not contain any desensitizing gum. The phosphate ion acts as the desensitizing material. The phosphate reacts rapidly with the aluminium plate surface.

Some of the advantages of using alkaline fountain solutions in newspaper offset printing are the following.

- There is no stripping of ink rollers.
- Blankets do not become glazed since the solution does not contain gum.
- The growth of fungus in the fountain pan is avoided.
- Linting (the phenomenon of accumulation of paper fibres on the rubber blanket that occur during lithographic printing, which is caused by paper with loose surface fibres and ink with high tack. This often leads to blinding of the image areas, due to the abrasive action of these fibres) is minimized, probably because it normally requires a little more fountain solution to keep the plate clean.
- Aluminium plates run clean and do not need to be gummed up, even after overnight process.

The alkaline fountain solutions may also cause some disadvantages such as foaming in the fountain solution, excessive emulsification of water in the ink and bleeding of ink pigments into the fountain solution, causing tinting on the sheets. With the use of alkaline fountain solutions the printing of ultraviolet inks has been found to be most successful.

Neutral Fountain Solutions

These are made up of neutral phosphates, or citrates, or salts of other weak acids. They can be used on commercial sheet-fed or web presses, but they are most frequently used in printing newspaper and business forms.

FUNCTIONS OF FOUNTAIN SOLUTIONS

The dampening of an offset plate is one of the most critical factors in the production of quality printing. To meet this requirement, fountain solutions are widely used in offset press. The main functions of fountain solutions are the following.

- To keep ink off the background (i.e., non-image area of the plate) with a film of water.
- To maintain the hydrophilic nature of the background.
- To quickly clean ink off the background during the press operation.
- To promote the fast spreading of water over the plate surface.
- To keep the flow of water evenly and smoothly through the dampening rollers on to the plate.
- To lubricate the plate and blanket.
- To control the excessive emulsification of ink and water.

MONITORING THE CONCENTRATION OF FOUNTAIN SOLUTIONS

An accurate control over the concentration (optimum) of a fountain solution is essential for consistent, high-quality results in offset printing. Low concentrations of fountain solution can cause drying of ink on the non-image area of the plate resulting in tinting, scumming, blanket piling, etc. High concentrations, on the other hand, bring about over-emulsification of the ink resulting in the weakening of colour strength and change the ink rheology (body and flow properties). The optimum concentration will lead to the proper wetting of the non-image areas of the plate. The two important key properties of a fountain solution are its pH and conductivity. The concentration of the fountain solution can be maintained by measuring its pH and conductivity.

pH of Fountain Solution

The pH of fountain solution must be controlled not only during the initial mixing but also during the printing process. This is essential for maintaining high-quality, trouble-free printing. Most fountain solutions use acid compounds to enable the gum arabic to stay in solution so that it can adhere to the non-image areas of the plate. The optimum pH for most acid fountain solutions is between 4.0 and 5.0. When the solution becomes more alkaline, the gumming agent loses its ability to desensitize the non-image areas, resulting in scumming, in which the ink replaces the gum on the plate. Scumming may also occur if the solution becomes too acidic because it can affect the protective layer of the plate. The latter type of scumming generally appears darker and is not as evenly distributed as scumming resulting from excessive alkalinity.

Increased acidity also slows down or inhibits the drying of ink, and can cause plate blinding, where the image area becomes less receptive to ink thereby causing a "ghost-like" image.

Decreased acidity can prevent the ink from adhering to the inking rollers, resulting in stripping.

Buffering agents are used in fountain solution concentrates to stabilize the pH level of the working fountain solution. The buffering agents keep the pH stable during the course of printing. Many modern fountain solutions are pH stabilized (or buffered), and so only small changes in pH are seen even when drastic changes occur in solution strength. Therefore, the pH measurement does not give an accurate indication of the fountain solution concentration. Today, conductivity testing is recognized as a much more accurate method.

Conductivity of Fountain Solution

Conductivity is the measurement of the ability of a fountain solution to conduct electricity. It is usually expressed in microsiemens (micromhos). In water or any solution, the conductivity increases directly with an increase in the amount of dissolved minerals. Hence, the conductivity of fountain solution increases directly with an increase in the amount of fountain concentrate used. Therefore, the conductivity is a better measure of the strength or concentration of fountain solution. Actually, pure water is a poor conductor of electricity. Partially ionizable substances such as gum arabic and alcohol are poor electrical conductors. They usually decrease the conductivity of fountain solutions. If the conductivity of different amounts of fountain concentrate in water is known, it is easy to measure the strength of a fountain solution by measuring its conductivity. To measure conductivity, an electronic conductivity meter is used. Many are combined with pH meters so that the printers can measure the pH and conductivity at the same time. The following procedure can be used to estimate the concentration of fountain solutions based on conductivity measurements.

1. Measure the conductivity and pH of the water that are used to prepare the fountain solution.

2. Add 1 ounce (29.6 ml) of fountain solution concentrate to a gallon (3.8 L) of water and remeasure both the conductivity and pH.

3. Add another ounce of fountain solution concentrate and measure again both the conductivity and pH. Repeat this procedure until the amount of fountain solution added exceeds the manufacturers' recommendations.

4. Plot a graph (Figure 15.1) with concentration values on the horizontal axis, and conductivity and pH values on the vertical axis.

The above procedure can be repeated and a similar graph can be developed if isopropyl alcohol or an alcohol substitute is also used in the fountain solution.

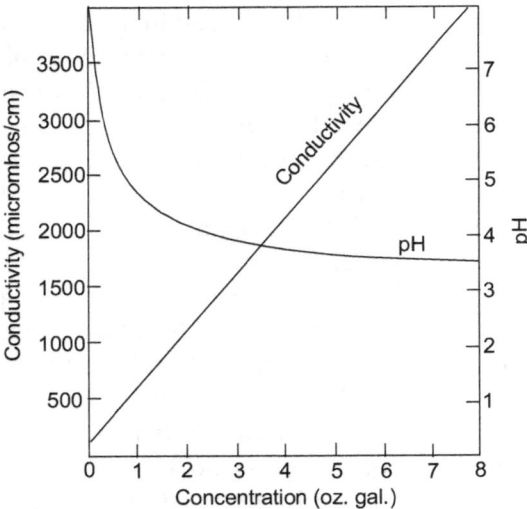

Figure 15.1 A typical graph of concentration versus conductivity and pH

It is clear from the above graph that pH remains constant from 3.5 to 8 ounces per gallon. Thus monitoring only the pH level through the press run would be useless in determining actual fountain concentration. Conductivity measurement, however, shows a linear variation with respect to concentration.

New graph must be made whenever the sample of water or the fountain solution concentrate changes. If the conductivity of the fountain solution is known, the amount of fountain concentrate to be added can be read directly from the graph.

Normally the conductivity of a fountain solution increases during the press run, because materials from the ink and paper may contaminate the fountain solution. These contaminates can interfere with the fountain solution materials and results in effective non-image protection which results in scumming, toning, tinting, etc. Therefore, conductivity measurements should be made before the fountain solution is used on the press and they should be checked at least twice a shift to determine if contamination requires draining tanks.

Hard water contains many dissolved minerals, which can increase conductivity. Water softeners are useful in controlling the mineral contents, which may help to ensure consistent water quality. Reverse osmosis and deionization units can be used to remove salts, minerals and some organic minerals from the water.

REVIEW QUESTIONS

1.   What are the ingredients of a fountain solution? Explain the function of each.

2.   Discuss the main functions of a fountain solution in offset printing.

3. How do the fountain solutions desensitize the non-image area of the lithoplate?

4. Write notes on:
   i. Acid fountain solution
   ii. Alkaline fountain solution

5. What are the advantages of using alkaline fountain solution in newspaper offset printing?

6. Discuss the importance of the following properties of fountain solutions.
   i. pH
   ii. Conductivity

7. 'Conductivity is a better measure of the strength or concentration of fountain solution'. Comment on the statement.

8. Describe the method of determining the fountain solution concentration based on conductivity measurements.

9. What are the adverse effects of variation of pH of fountain solution? How is the pH of fountain solution stabilized?

10. Explain the function of isopropyl alcohol in fountain solutions.

11. What are the disadvantages of using isopropyl alcohol in fountain solution?

12. Name the alcohol substitutes that are suitably used in place of isopropyl alcohol in fountain solutions.

ADDITIONAL READING

De Jidas, L.P. (1992). "Alcohol Substitutes –Making them Work for You." *GAFT World*. Vol.4(1). p. 21.

MacPhee, J. (1990). "Tips on How to Maximize the Performance of Dampening Systems". *GAFT World*. Vol. 2(5). p. 35.

GLOSSARY

Abrasion resistance   The ability to withstand the effects of repeated rubbing and scuffing.

Abrasive paper   The paper covered on one or both sides with abrasive powder, e.g. emery, sandpaper, etc.

Additive pre-sensitized plates   The pre-sensitized plates in which a wear-resistance lacquer or pigmented resin is added to the image areas during plate development.

Adhesion   The physical attraction of the surface of one material to the surface of another.

Agar–Agar   A heteropolysaccharide which is extracted from certain East Indian seaweed. Solutions of agar in hot water set to a gel when they cool.

Alkali   A base which is soluble in water, e.g. NaOH, KOH.

Angstrom unit (Å)   A non-SI unit of length used to express wavelengths of light, bond lengths, and molecular sizes. $1 \text{ Å} = 10^{-10}$ m $= 10^{-8}$ cm.

Anilox roller   An engraved steel roller used to meter ink to the plate cylinder of a flexopress.

Anion   A negative ion; an atom or group of atoms that has gained one or more electrons.

Bagasse   Paper-making fibre made from the remaining residue after sugar cane has been crushed and its juices have been extracted.

Bag paper   Any paper made to be used in the manufacturing of bags.

Bamboo   A woody plant of grass family whose fibre is used for paper-making in some countries.

Bible paper   Light, thin, strong, opaque paper used for printing bibles, dictionaries, sales manuals or other jobs where bulk may be a problem. Not suitable for writing because of its absorbency.

Blanc fixe   Barium sulphate, a white semi-opaque pigment.

Blanket   A multilayered composite of rubber and fabric, which is used in the lithographic process to transfer the image from the plate to the paper.

Blend   A mixture, such as a mixture of solvents or a mixture of inks.

Blotting paper   An unsized paper used generally to absorb excess ink from freshly written manuscripts, letters and signatures.

Blueprint   An image formed by white lines on a blue background and made by contacting pen or pencil originals that were made on vellum (a fine paper resembling parchment) or other translucent materials. They are useful for reproducing production drawings, construction plans and architectural drawings.

Blueprint paper   Direct-copy-process paper manufactured with high rag content (fibrous content in cotton) to produce a smooth finish, a high wet strength and a high degree of absorbency. To make a copy, the original or translucent master is exposed in direct contact with the blueprint paper.

Board   A heavy paperlike material, such as that used for mounting art or making displays.

Bond paper   A grade of writing or printing paper, where strength, durability and permanence are essential requirements; used for letterheads, business forms, etc.

Boric acid H_3BO_3, used as buffering agent.

Boiling point   The temperature at which the vapour pressure of a liquid is equal to the applied pressure.

Boxboard   Paperboard of sufficient caliper and strength to be used in the manufacture of boxes.

Brass   An alloy of copper containing 20–40 % zinc.

Brightness   Quality of whiteness intensity emitted from printed or unprinted surfaces.

Bronze   An alloy of copper containing 10–25% tin.

Buffer range   The pH range at which a buffer is effective in its action.

Bulk polymerization   The simplest polymerization technique which involves the use of monomer and initiator as the main components; the reaction takes place in a homogeneous system.

Bulky paper   Grades of paper made predominantly from mechanical pulp to a specific and high bulk.

Carbohydrates   The complex non-nitrogenous compounds composed of carbon, hydrogen and oxygen only.

Carton board   A rigid wood-fibre-based packaging material. Carton-board is usually of at least 180 g/m^2 basis weight and 250 microns thickness.

Cation   A positive ion; an atom or group of atoms that has lost one or more electrons.

Cellulose acetate   A transparent thermoplastic material prepared from cellulose by treatment with acetic anhydride.

Cellulose nitrate   A transparent thermoplastic material produced by nitrating cellulose with a mixture of nitric acid and sulphuric acid.

Chlorination   The introduction of chlorine into molecules of inorganic and organic substances.

Chemical resistance   The ability of a material to resist the action of common chemicals.

Chroma One of the attributes of colour, characterized by its purity or saturation (or strength).

Cigarette paper A light-weight, unsized paper (grammage 18 to 24 g/m^2) used in making cigarette. It contains about 30% calcium carbonate as filler to control the burning rate and match it with tobacco burning rate. Very long fibres such as jute, cotton, etc. are used to achieve high strength and porosity.

Coated paper A paper coated with clay or other white pigments and a suitable binder to produce a smooth finish.

Cobb test A method for measuring the degree of sizing of paper or paperboard by determining the weight of water absorbed under prescribed conditions during a specified time period.

Collagen The major protein component of connective tissue.

Cold-set ink Solid inks which must be melted and applied on a press. They consist of pigments dispersed in plasticized waxes.

Colligative properties Physical properties of solutions that depend upon the number but not the kind of solute particles present (such as osmotic pressure, boiling point elevation, freezing point depression).

Colloidal state A substance is said to be in the colloidal state, when it is dispersed in another medium in the form of very small particles having size range 10–2000 Å (1–200 mμ).

Colourant The material may be a pigment or a dye or a combination of the two, which impart colour to printing inks.

Colour strength In printing inks, it means the effective concentration of colouring material per unit weight or volume.

Contact print A photographic copy made by exposing a negative or positive in direct contact with the sensitive material (paper).

Continuous tone A photographic image which contains gradient tones from black to white.

Corrugating medium A type of paperboard made from unbleached, semi-chemical pulp, especially NSSC (neutral sulphite semi-chemical) hard wood pulp, and recycled fibre from corrugated boxes. It is formed into a fluted (wavy) structure and sandwiched between two plies of linerboard (unbleached Kraft softwood sheet) to form the corrugated structure for boxes.

Cover paper A wide variety of fairly heavy plain papers which are converted into covers for books, catalogues, brochures, pamphlets, etc. Good folding qualities, printability, and durability characterize it.

Cross direction The direction perpendicular to the direction of web travel through a paper machine or through a web press.

Curing   The chemical conversion of a wet coating or printing ink film to a solid film.

Currency paper   The paper used for printing currency. These papers have very high folding endurance, permanency, tensile strength and suitable for four-colour printing, with watermark and other falsification safeguards such as embedded metal strip. Often contains cotton fibres.

Dampening roller   A roller that transfers dampening solution to the plate during printing.

Debarking   Removing the bark from wood logs prior to conversion to mechanical or chemical pulp.

Deep-etch image   The image area of an exposed lithoplate which is deep-etched by treatment with a deep-etch solution.

Delignification   The process of removal of lignin from wood pulp.

Desalination   The process of removal of dissolved salts from seawater.

Desensitize   The treatment on a lithographic plate so as to make the non-image areas as water-receptive and ink-repellent.

Diarylide yellow   A strong yellow toner used in many types of printing inks.

Diazo compound   A mixture of diazonium salts with an azo dyestuff component. The salts are sensitive to ultraviolet, violet and blue light.

Diazonium salt   A diazonium salt is a compound with general formula $Ar—N{=}N^+X^-$, where Ar represents a substituted benzene ring and X^- is a halide ion such as chloride. They are used to manufacture many different organic compounds, including azo dyes.

Dichromated colloid   A combination of a dichromate and a natural or synthetic polymer, such as potassium dichromate and gelatin; ammonium dichromate and polyvinyl alcohol.

Diluent   A solvent that is added to reduce the viscosity.

Dimensional stability   The property of a material to resist the dimensional changes with the change in its environmental conditions.

Diisocyanate   Organic compounds having two $—N{=}C{=}O$ groups per molecule.

Dioxan   An organic solvent for cellulose acetate, resins and rubber materials ($C_4H_8O_2$).

Dissociation constant   An equilibrium constant that applies to the dissociation of a complex ion into a simple ion and coordinating species (ligands).

Distillation   The separation of a liquid mixture into its components on the basis of differences in boiling points.

Distributing roller   A rubber covered roller which conveys ink from the fountain on to the ink drum.

Double decomposition A reaction in solution in which two compounds take part; both are decomposed, and by an exchange of radicals two new compounds are formed.

Dot gain A printing defect in which dots print larger than they should, causing darker tones or colours.

Drying oil The oils which possess the property of hardening to a tough film by oxidation and polymerization.

Ductor roller An ink or water roller that alternatively contacts the fountain roller to pick up a load of ink or water and then transfers it to a distributor roller.

Duplicate printing plate The relief plate made from a master plate that was not intended to be locked up in the printing form.

Durometer An instrument used for measuring the hardness of printing roller.

Elasticity (resilience) The property of a material (such as rubber plates, blankets, rollers), that enables it to return to its original size and shape after being stretched or deformed.

Electrical insulating paper Papers used for electrical insulation purposes. Strong, pinhole free paper, sometimes impregnated with synthetic resins and made from unbleached kraft pulp. Electrical insulating paper must neither contain fillers nor conductive contaminants (metals, coal, etc.) nor salts or acids.

Electron beam A stream of high-velocity electrons.

EB ink The printing ink that dries by the application of electron beam.

Electronegativity The tendency of an atom in a molecule to attract the shared pair of electrons to itself.

Electroplating The process in which a metal is coated on the surface of a metallic object by passing direct current through an electrolytic solution.

Electrostatic printing A copying or duplicating process in which an image is created on an electrically charged carrier by light reflecting from the copy to be reproduced.

Electrotyping The process of electrolytic production of electrotype plates (i.e., duplicate printing plates).

Emulsification The dispersion of one liquid into another immiscible liquid in the form of fine droplets.

Emulsifier Substances used to facilitate the formation of emulsions and to improve their stability.

Emulsion polymerization A technique in which a stabilized emulsion of the monomer in water is subjected to polymerization reaction.

Equilibrium A reversible reaction in which the rate of the forward reaction is equal to the rate of reverse reaction.

Etching   The process of removal of an unwanted metal from a plate or cylinder with an acid.

Exposure   The step in photographic processes during which light produces the image on the light-sensitive coating.

Exothermic   A reaction in which heat is given out, i.e., the temperature rises as the reaction proceeds.

Fatty acid   A long-chain aliphatic acid.

Felt side   The top (smoother) side of the sheet in paper manufacturing, opposite the wire side. The usual side used for printing.

Ferrochrome   Alloy of iron with chromium.

Filter paper   Unsized paper made from chemical pulp, in some cases also with an admixture of rags, sometimes with a wet-strength finish. Filtration rate and selectivity, which are both dependent on the number and the size of the pores, can be controlled by specific grinding of the pulps and creeping.

Flexography   A relief printing process in which an image is transferred from rubber or photopolymer plates, using liquid inks, to a variety of substrates.

Flexographic ink   A solvent type, low-viscosity, fast drying ink that contains both dyes and pigments to use with rubber and photopolymer plates for flexographic printing.

Foam   A colloidal dispersion of a gas in a liquid.

Fog   A visible aggregate of minute water droplets or ice crystals suspended in the atmosphere.

Form rollers   The last group of rollers in the ink distribution system of a printing press that come in direct contact with the printing plate and transfer ink to the plate.

Free radical   An atomic or molecular species having an odd or unpaired electron. They are the highly active species.

Fountain/fountain pan   An ink reservoir on a printing press, which includes controls for the uniform removal and distribution of ink to distributor rollers, the plate cylinder, or anilox ink rollers, depending on the printing process being used.

Fountain solution   In lithography, a solution used for preventing the non-printing areas of the plate from accepting ink.

Fuming sulphuric acid (oleum)   A solution of sulphur trioxide (about 20%) in sulphuric acid, $(H_2S_2O_7)$.

Galvanizing   The process of coating a base metal (iron / steel) surface with zinc metal.

German silver   Alloy of copper containing 30% nickel and 20% zinc.

Ghost-like image   A design of faint image in areas which are not intended to receive that portion of the image.

Glass electrode It consists of a thin-walled glass bulb (with high electrical conductance) filled with 0.1 M HCl and inserted with a Ag–AgCl electrode. It is a pH-sensitive electrode and hence used in pH determination.

Glassine paper A transluscent paper made from highly beaten chemical pulp and subsequently supercalendered.

Gloss The reflectance of light from a surface.

Gloss ink An ink that dries with a minimum of penetration into the stock and gives high lustre.

Grain Arrangement or direction of fibres in a fibrous material such as paper.

Gravure A type of printing using intaglio process, i.e , the ink is placed in cells below the plate surface.

Gravure ink A quick-drying, low-viscosity ink based on volatile solvents, which is used in gravure printing.

Greasing *See* Scumming.

Gum/Gumming agent A water-soluble colloid (such as gum arabic) used for coating a lithographic plate to make the non-image areas ink-receptive and to preserve the plate for future use.

Half-tone A printed image composed of dots of varying frequency (number per sq. in.), size or shape, producing tonal gradations. The term is also applied to the process, plates and ink used to produce this image.

Hansa yellow A light-resistant organic yellow pigment.

Hardness The property possessed by a material which enables it to resist penetration or abrasion or scratching by other materials.

Headbox The part of paper-making machine whose primary purpose is to deliver a uniform dispersion of fibres in water at the required rate through the opening to the paper machine wire.

Heat-set ink A special ink composed of pigment, litho varnish, solvent (heat-set oil) and modifier (wax compound). They dry under the action of heat by the evaporation of their high boiling solvent.

Hemicellulose A constituent of wood that, like cellulose, is a polysaccharide but is less complex and easily hydrolysed.

Heterogeneous system A system comprising several homogeneous parts (phases) separated by interfaces.

High contrast In photography, a reproduction with high gamma in which the difference in darkness (density) between neighbouring areas is greater than the original.

Hue One of the attributes of colour, which is determined by its dominant wavelength.

Hydrogen bond An electrostatic attraction that exists between hydrogen atom of one molecule and electronegative atom of either the same molecule (intra-molecular) or another molecule (intermolecular). Hydrogen bond is observed in molecules where hydrogen atom is bonded to highly electronegative atoms (such as fluorine, oxygen and nitrogen).

Hydrogenation The reaction in which hydrogen adds across a double or triple bond.

Hydrogenating catalyst The catalyst (such as finely divided nickel) which is used in hydrogenation process (such as in hydrogenation of oils).

Hydrolysis The reaction of a substance with water or its ions.

Hydrophilic (water-receptive) Receptive to water and fountain solution, rather than to oils and inks.

Hydrophobic (water repellent) Receptive to oils and inks, rather than to water and fountain solution.

Hygroscopic The property of substances which makes them to absorb moisture from the air.

Hypo Sodium thiosulphate or sodium hyposulphite, a chemical used to fix the image on a photographic film after it has been developed.

Hickies Defects in a print appearing as specks surrounded by an unprinted 'halo' caused by a foreign object on the plate or blanket.

Indicator A compound which, by changing colour, shows whether a solution is acidic or alkaline; or which indicates the end point of an acid–base titration.

Initiators The catalysts (thermally unstable peroxides such hydrogen peroxide, acetyl peroxide, benzoyl peroxide) used in the free radical polymerization.

Infrared Electromagnetic radiation with wavelength range 780 nm to 1 mm.

Injection moulding The method of fabrication of thermoplastics into desired shaped articles.

Ink fountain In printing presses, the device which stores and supplies ink to the inking rollers.

Inkometer An instrument used for measuring the tack of printing inks.

Ink receptivity Uniform accepting of ink on paper surface.

Ink roller The roller which transfers ink to plate during printing.

Ink setting The increase in viscosity that occurs immediately after ink has been transferred to a substrate.

Interfacial tension The force acting per unit length along the interface of two liquids.

Intermolecular forces The forces between individual particles (atoms, molecules, ions) of a substance.

Ion An atom or a group of atoms that carries an electric charge.

Ionic lattice A regular three-dimensional arrangement of ions in a crystal.

Ionization The process of spontaneous splitting up of an electrolyte into ions, when dissolved in a solvent (like water).

Isoelectric point Each protein has a characteristic isoelectric point at which its ionization and solubility is minimum.

Kaolin A fine white clay used as a filler or incorporated in the coating to serve as a pigment in the manufacture of paper.

Kinetic energy Energy that matter processes by virtue of its motion.

Kraft substrates The paper or other paper products made from kraft pulp.

Lacquering The application of a layer of lacquer on a printed sheet for protection or appearance (to improve gloss).

Laminate A product made by bonding together two or more layers of material(s), usually with an adhesive.

Laminated paper A paper built up to a desired thickness or a given desired surface by joining together two or more webs or sheets. The papers thus joined may be alike or different; a totally different material, such as foil, may be laminated with paper.

Le Chatelier's principle It states that a system at equilibrium, or striving to attain equilibrium, responds in such a way as to counteract any stress that is applied it.

Letterpress A process of typographic (raised type) printing using oil-based inks.

Lignin A polymeric, phenolic compound that serves to bind the cellulose fibres together.

Linseed oil Oil obtained by pressing flaxseed and is used as a raw material in the production of lithographic varnish.

Lithographic printing (lithography) A process of planographic printing involving two chemically different areas on the plate (one is receptive to ink and the other receptive to fountain solution).

Lithopone A white pigment, which is a mixture of zinc sulphide (ZnS, 28–30%) and barium sulphate (BaSO$_4$, 70–72%).

Long run A high printing number required for a job.

Machine direction The direction of paper parallel to its forward movement on the paper machine.

Machine finished (MF) Smooth paper calendered on the paper machine.

Machine glazed (MG) Paper with a glossy finish on one side produced on the paper machine by a Yankee cylinder.

Map paper The paper used for making maps must be subject to minimum change in dimensions with moisture to avoid poor register of colours. Wet strength properties are often demanded.

Matte finish A dull paper finish, without gloss or lustre.

Mechanical engraving A method of cutting an image into a gravure cylinder mechanically by using a lathe like machine on which the cylinder is mounted and is engraved with a special diamond-stylus engraving tool.

Melting point   The temperature at which a solid changes to a liquid. The temperature at which liquid and solid coexist in equilibrium.

Milori Blue   A green shade iron blue pigment. Also called as bronze blue, Chinese blue.

Mineral oils   Hydrocarbon oils of high molecular weight obtained from petroleum.

Metallic inks   They consist of a suspension of fine metal particles in a vehicle that serves to bind the powder to the surface being printed.

Microorganism   Any organism which can be observed only with the help of a microscope (include bacteria, virus, fungi, etc.).

Microwave   Electromagnetic radiation with wavelength range 1mm to 1m.

Molar concentration   The concentration expressed in molarity. A molar solution contains the gram molecular weight of the compound dissolved in one litre of solution.

Mole   An amount of substance equal in grams to the sum of the atomic weights.

Molecular formula   The formula that indicates the actual number of atoms present in a molecule of a molecular substance.

Mordant   The substance which provides a chemical link or bridge between the dye and the fibre (such as tannic acid, albumin, salts of aluminium, chromium, iron, etc.).

Negative   In photography, film containing an image in which the values of the original are reversed so that the dark areas appear light and vice versa (*see* Positive).

Newsprint   A low-cost paper made from ground wood or mechanical pulp, with some chemical pulp added for strength. The paper is not sized and is very absorbent, which permits quick absorption of the news inks.

Nickel steel   Steels containing nickel.

Nichrome   Nickel alloy containing 20% chromium.

Non-piling agent   An additive, in fountain solution, that eliminates any chances of piling.

Oleophilic   Ink-receptive.

Oligomer   A polymer molecule made up of only a few monomer units (polymer with a very low degree of polymerization).

Offset printing   Indirect form of printing in which the ink is transferred from the printing plate to a rubber blanket and subsequently to the substrate.

Osmosis   The process by which solvent molecules pass through a semipermeable membrane from a dilute solution into a more concentrated solution.

Osmotic pressure The hydro-static pressure produced on the surface of a semipermeable membrane by osmosis.

Oxidation The chemical reaction which involves one of the following.

- addition of oxygen to an element or compound, or
- removal of hydrogen from a compound, or
- loss of electrons.

Oxidizing agent A substance that oxidizes another substance and itself gets reduced.

Packaging paper A paper or paperboard used for wrapping or packing goods.

Paperboard A heavy weight, thick, rigid and single or multi-layer sheet. What differentiates paperboard from paper is the weight of the sheet. If paperboard is very heavy it is called board.

Passive The metals which quickly develop a protective oxide film on exposure to air or under the action of oxidants (such as nitric acid).

Parts per million (ppm) Con-centration expressed as parts of solute per million parts of solution.

Permanence The resistance of paper to change in one or more of its characteristics during storage or aging.

Petroleum fractionation The process of separation of petroleum (crude oil) into different useful fractions on the basis of their boiling points.

Petroleum naphtha An aliphatic hydrocarbon solvent derived from petroleum.

pH meter A digital instrument for the measurement of pH of an unknown solution.

Phosphoric acid H_3PO_4, orthophosphoric acid.

Photoengraving The method of making relief plates by coating a metal surface (copper, zinc, magnesium) with a photosensitive coating.

Photoengraving glue Dichro-mated fish glue used as a light-sensitive coating in photo-engraving.

Photographic paper The base paper used for the production of photographic papers is a dimensionally stable, chemically neutral chemical pulp paper with wet strength properties that must be free from contaminants. Today papers are coated on both sides with a thin polyethylene film. The cooking prevents chemicals and water entering the paper during development. This also permits shorter rinsing and drying cycles.

Photoinitiator A substance which, by absorbing light, undergoes splitting to form free radicals and initiates the polymerization reaction.

Photon A discrete packet of energy associated with electro-magnetic radiation. Each photon carries energy E proportional to the frequency v, of the radiation $E = hv$, where h is Planck's constant.

Photosensitizer A substance which, by absorbing light, passes its energy to another substance which then reacts.

pi bond (π bond) In the valence bond theory, a pi bond is a valence bond formed by side-by-side overlap of p orbital on two bonded atoms.

Picking The lifting of areas of paper surface during printing which happens when ink tack is stronger than the surface strength.

Pigment flocculation The aggregation of pigment particles in ink, creating clusters or chains, which may result in a loss of colour strength and change of hue.

Piling An accumulation of ink or other materials on a blanket in sufficient quantity to affect the quality of print in the image or non-image areas.

Plastic binding A type of mechanical binding using a plastic centre strip from which curving prongs extend.

Plasticizers The liquid or solid additives used to impart flexibility.

Plastisol A suspension of powdered resin in a plasticizer.

Polysaccharide A complex carbohydrate whose molecules contain a large number of monosaccharide residues, e.g. cellulose, starch, etc.

Positive In photography, film containing an image in which the dark and light values are the same as the original.

Pre-sensitized litho plate Lithographic plate precoated with a light-sensitive coating.

Press run The total number of copies of a job printed to fully or partially satisfy the conditions of the client.

Printing block A block carrying a design by which the ink is ultimately transferred directly or indirectly to the substrate.

Print quality In paper, the properties of paper that affect its appearance and the quality of reproduction.

Prussian Blue A red-shade iron blue pigment.

Publication paper On-machine-coated printing paper suitable for colour printing. They include: wood-free printing and writing paper and ivory wood-free printing and writing paper.

Pumice stone A highly porous igneous rock usually containing SiO_2 (67–75%) and Al_2O_3 (10–20%).

Quick-set inks These inks dry or set within a few seconds by filtration, coagulation, selective absorption and oxidation. The vehicles are generally special resin–oil combinations which, after the ink has been printed, separate into a solid material which remains on the surface as dry film and an oily material which penetrate rapidly into the stock.

Record paper (Ledger paper) A smooth-finished paper used for bookkeeping or any other purpose where strength, absence of glare, suitability to writing with ink and irascibility are required.

Redox reaction The reaction in which both oxidation and reduction occur.

Reduction The opposite of oxidation (*see* Oxidation).

Reducing agent The substance that reduces another substance and is oxidized.

Refractive index The ratio of the speed of light in vacuum to the speed of light through a material.

Relative humidity The ratio of the amount of water vapour actually present in the air to the amount of water vapour necessary for saturation at the existing temperature.

Reversible reaction The reaction that does not go to completion and occurs in both the forward and reverse directions.

Reverse osmosis The phenomenon of forcing of solvent molecules to flow through a semipermable membrane from a concentrated solution into a dilute solution by the application of greater hydrostatic pressure on concentrated side than the osmotic pressure opposing it.

Relief A raised printing surface such as that used in letterpress or flexographic printing.

Reference electrode An electrode having a known electrode potential that remains constant and with reference to which the electrode potential of other electrodes can be measured.

Rosin A natural resin obtained from the fluid 'oleoresin', which is tapped from pine trees.

Rosin oil Oil obtained from the destructive distillation of rosin.

Rubometer An instrument used to measure the resistance of ink to rubbing, scuffing and scratching.

Runnability The property of paper that enables the press to feed the sheet mechanically through the press without difficulty.

Sand blasting A surface-cleaning method, in which sand is introduced into an air stream under high pressure and the blast is impacted on the metal surface to be cleaned. It removes the oxide film or scale and other inorganic deposits from the metal surface.

Saturated calomel electrode A secondary reference electrode with potential 0.2422 V at 25°C. It consists of a glass tube containing mercury and mercurous chloride (calomel, Hg_2Cl_2) and filled with saturated KCl solution.

Saturated hydrocarbons Hydrocarbons that contain only single bonds. They are also called alkanes or paraffins.

Saturated solution The solution in which no more solute will dissolve.

Scumming   In litho-plate making, an unfavourable condition that causes the plate to pick up ink in the non-image areas. (This is because the non-image areas of the plate were not completely desensitized.) This condition is sometimes called 'greasing'.

Screen-printing   A process in which an image is transferred to a substrate by squeezing the ink through the unblocked areas of a metal or fibre screen.

Semi-permeable membrane   A thin partition between two solutions through which certain molecules can pass but others cannot.

Show through (See through)   The degree to which printing on the reverse side of a sheet can be seen through the sheet under normal lighting conditions. This effect is due largely to a lack of insufficient opacity in the sheet.

Smoke   A colloidal dispersion of a solid (such as carbon particles) in gas or air.

Snow-flaky print   Very small white specks that appear on a printed offset sheet, caused by water droplets that remain in the printing nip during image transfer.

Softening point   The temperature at which a plastic material will begin to deform with no externally applied load.

Solder   A low-melting alloy of tin and lead, which is used in soldering.

Solvation   The process by which solvent molecules surround and interact with solute ions or molecules.

Solvent   The dispersing medium of a solution.

Solvent cleaning   The method of removal of organic impurities (such as oil, grease, etc.) from a metal surface on treatment with an organic solvent (such as carbon tetrachloride, toluene, xylene, etc.).

Soya bean oil   A semi-drying vegetable oil used in the preparation of printing ink vehicles.

Specific gravity   It is the ratio of the mass of a substance to the mass of an equal volume of water.

Starch   A complex carbohydrate of high molecular weight, found in most vegetables and cereals. Represented by the general formula $(C_6H_{10}O_5)_n$.

Stabilizing agent   Substances which are used to prevent the coagulation of colloidal dispersions.

Stainless steel   Steel containing more than 12% chromium, which is corrosion resistant.

Stock   Paper, paperboard or other paper products on which an image is printed, copied or duplicated.

Strike-through   The penetration of the vehicle of a printing ink through the sheet so that the ink is apparent on the opposite side.

Stripping A condition in which the ink fails to adhere to and distribute uniformly on the metal rollers of the press.

Structural isomers The compounds having same molecular formula but different structural formula.

Substrate Any material upon whose surface an image can be applied or transferred, a coating can be applied, to which another material can be laminated, or that can be used as the base for any number of operations in the graphic arts industry.

Surface tension The force (in Newton) acting at right angles to the surface of a liquid along unit length (1 metre) of the surface. The force, that tends to contract the surface of a liquid.

Sugar ($C_{12}H_{22}O_{11}$; cane sugar) A carbohydrate of disaccharide series.

Tall oil A mixture of fatty and rosin acids obtained as a by-product of a sulphate paper-making process and used in the preparation of printing ink vehicle.

Tanning The process of making leather from hides under the action of tanning agents (such as formaldehyde, alum etc.).

TAPPI The Technical Association of the Pulp and Paper Industry, an international organization of professional people from the paper and allied industries whose stated purpose is to conduct research and to establish technical standards and testing procedures regarding the manufacture and use of pulp and paper.

Tensile strength The property of a material that enables it to withstand a force acting upon it with a tendency to break it by tearing.

Tinting A uniform discoloration of the background caused (in lithography) by the bleeding of the pigment in the fountain solution.

Tissue A fine, thin paper used for a variety of purposes where a delicate, light-weight paper is required. Normally a paper sheet weighing less than 40 grams per metre square is called tissue.

Toner The imaging material (a dry powder that contains a pigment, usually carbon, and a thermoplastic resin) used in electrostatic printing. In inks, a highly concentrated pigment and/ or dye, used to modify the hue or colour strength of an ink.

Toxicity The ability of a substance to cause injury to a living tissue.

Tracing cloth A duplicate drawing with a black image made on blue or white waterproof cloth. It is probably the longest-lasting type of duplicate.

Tracing paper A transparent paper manufactured for tracing.

Transparent ink Inks which lack hiding power and permit transmission of light, thus allowing previous printing or substrates to show through.

Trapping The property of an ink that facilitates its transfer equally to an unprinted substrate and to a previously printed ink film.

Two-sidedness In paper, the property denoting difference in appearance and printability between its top (felt) and wire sides.

Type metal Alloy of lead (75%) and antimony (20%).

Tung (Chinawood) oil A vegetable oil having exceptional drying properties and chemical resistance.

Ultraviolet(UV) light An electromagnetic radiation with wavelength in the range 100 to 380 nm.

Unsaturated hydrocarbons Hydrocarbons that contain double or triple carbon–carbon bonds.

UV ink Ink drying by the application of ultraviolet radiation.

van der Waals force A weak inter-atomic or intermolecular attraction existing between atoms or molecules of a substance without the formation of covalent or ionic bond.

Vapour degreasing A more effective method of solvent cleaning, in which the solvent (such as trichloroethylene) is vaporized by heating and the vapours are made to condense on the metal surface to be cleaned. The condensed liquid dissolves and washes away the oil, grease and other organic matters.

Varnishing The application of an overprint varnish to the surface of a printed job to protect it or to increase its gloss.

Viscosity The property of a material by virtue of which it tends to resist deformation or flow.

Volatile solvent The solvent which readily vaporizes at room temperature.

Wallboard A paperboard used for wallcovering.

Water-based inks Inks containing a vehicle whose binder is water-soluble or water-dispersible.

Watercolour inks Inks generally employed in the printing of wall paper, greeting cards and novelties. They are based on a vehicle composed essentially of gum arabic, dextrin, glycerine and water.

Waxing The application of a layer of wax to the reverse side of galleys, pages or repro copy to form a pressure-sensitive adhesive used to paste up mechanicals for reproduction.

Waterless lithography A plano-graphic process like lithography but it prints without water.

Weak acid/base The acid or base which undergoes partial ionization in solution.

Wear resistance The resistance offered by a material against wear and tear.

Weather resistance The ability of a material to withstand the change in the weather conditions.

Wet end The end of paper-making machine, where the fibres are formed into paper, located between the headbox and the drier section.

Web   A roll of printing paper cut from the paper machine log to the desired width and wound on a core to the diameter requested by the printer.

Web printing   Offset printing on a press designed to use paper supplied in rolls.

White light   An electromagnetic radiation with wavelength in range, 400–800 nm or 4000–8000 Å.

Wire side   The side (usually rough) of sheet which is in contact with the paper machine wire during paper manufacture; the side opposite to the felt side.

X-rays   An electromagnetic radiation with wavelength between 0.01–100Å.

INDEX